Sex and the Developing Brain

Colloquium Series on The Developing Brain

Editor
Margaret McCarthy, Ph.D.
Professor and Associate Dean for Graduate Studies
Department of Physiology
University of Maryland School of Medicine
Baltimore, Maryland

The goal of this series is to provide a comprehensive state-of-the-art overview of how the brain develops and those processes that affect it. Topics range from the fundamentals of axonal guidance and synaptogenesis prenatally to the influence of hormones, sex, stress, maternal care, and injury during the early postnatal period to an additional critical period at puberty. Easily accessible expert reviews combine analyses of detailed cellular mechanisms with interpretations of significance and broader impact of the topic area on the field of neuroscience and the understanding of brain and behavior.

My research program focuses on the influence of steroid hormones on the developing brain. During perinatal life, there is a sensitive period for hormone exposure during which permanent cytoarchitectural changes are established. Males and females are exposed to different hormonal milieus and this results in sex differences in the brain. These differences include alterations in the volumes of particular brain nuclei and patterns of synaptic connectivity. The mechanisms by which sexually dimorphic structures are formed in the brain remains poorly understood.

I received my PhD in Behavioral and Neural Sciences from the Institute of Animal Behavior at Rutgers University in Newark, NJ in 1989. I then spent three years as a post-doctoral fellow at the Rockefeller University in New York, NY and one year as a National Research Council Fellow at the National Institutes of Health, before joining the faculty at the University of Maryland. I am a member of the University of Maryland Graduate School and the Center for Studies in Reproduction. I am also a member of the Society for Behavioral Neuroendocrinology, the Society for Neuroscience, the American Physiological Association, and the Endocrine Society.

Sex and the Developing Brain
Margaret M. McCarthy
www.morganclaypool.com

ISBN: 9781615040605   paperback

ISBN: 9781615040612   ebook

DOI: 10.4199/C00018ED1V01Y201010DBR001

A Publication in the Morgan & Claypool Life Sciences series

THE DEVELOPING BRAIN

Book #1

Series Editor: Margaret M. McCarthy, University of Maryland

Series ISSN Pending

Sex and the Developing Brain

Margaret M. McCarthy
University of Maryland

THE DEVELOPING BRAIN #1

MORGAN&CLAYPOOL LIFE SCIENCES

ABSTRACT

The brains of males and females, men and women, are different—that is a fact. What is debated is how different and how important those differences are. Sex differences in the brain are determined by genetics, hormones, and experience, which in humans includes culture, society, and parental and peer expectations. The importance of nonbiological variables to sex differences in humans is paramount, making it difficult if not impossible to parse out those contributions that are truly biological. The study of animals provides us the opportunity to understand the magnitude and scope of biologically based sex differences in the brain and understanding the cellular mechanisms provides us insight into novel sources of brain plasticity. Many sex differences are established during a developmental sensitive window by differences in the hormonal milieu of males versus females. The neonatal testis produces large amounts of testosterone, which gains access to the brain and is further metabolized into active androgens and estrogens, which modify brain development. Major parameters that are influenced by hormones include neurogenesis, cell death, neurochemical phenotype, axonal and dendritic growth, and synaptogenesis. Variance in these parameters results in sex differences in the size of particular brain regions, the projections between brain regions, and the number and type of synapses within particular brain regions. The cellular mechanisms are both region and endpoint specific and invoke many surprising systems such as prostaglandins, endocannabinoids, and cell death proteins. By understanding when, why, and how sex differences in the brain are established, we may also learn the source of strong gender biases in the relative risk and severity of numerous neurological diseases and disorders of mental health, including but not limited to autism, dyslexia, attention deficit disorder, schizophrenia, Alzheimer's, multiple sclerosis, Parkinson's, and major depressive disorder.

KEYWORDS

androgen, estrogen, hypothalamus, preoptic area, hippocampus, sensitive period, synaptogenesis, neurogenesis, apoptosis

Contents

Introduction 1

Sex Determination versus Sex Differentiation 3

Masculinization, Feminization, and Defeminization 7

Steroid Hormones Are Potent Modulators of Brain Development 9

Sex Differences in the Brain Are Established During a
Developmental Sensitive Window 15

   Steroid Levels in the Developing Brain 15

   Organizational/Activational Hormone Effects on the Brain 18

Sex Differences in Physiology and Behavior Are Coordinated 19

   Ovulation Begins in the Brain 20

   Female Sex Behavior Is Coordinated with Ovulation 24

   Male Physiology and Behavior Are Not Temporally Constrained 26

   Changes in the Brain Induced by Steroids during Development Direct Adult
      Physiology and Behavior 27

Knockouts of the Rule: Mice with Null Mutations of Steroid Receptors,
Steroidogenic Enzymes, and Binding Proteins 29

Steroids Influence Multiple Endpoints to Organize the Brain 31

   Steroids Organize the Developing Brain by Altering Cell Survival 31

   Steroids Organize the Brain by Altering Cell Proliferation 33

Neuronal Migration Is Not Strongly Regulated by Steroids 35

Steroids Regulate Trophic Factors and Activity-Dependent Survival 35

Steroids Impact on Axonal Projections, Dendritic Branching, and Connections 37

Steroids Organize the Developing Brain by Altering Synaptic Connectivity 39

Steroids Organize the Developing Brain by Altering Neurochemical Phenotype 43

Vasopressin Is a Model of Steroid-Mediated
Sexual Differentiation of the Brain 45

Vasopressin Demonstrates Unique Parameters Associated with
Sexual Differentiation of the Brain 45

Cellular Mechanisms of Steroid-Mediated Organization of the Brain 47

Prostaglandins Masculinize the Preoptic Area and Sexual Behavior 47

GABA Induces Sex Differences in Astrocytes in the Arcuate Nucleus 52

Glutamate Release Is Critical to Sex Differences in Synaptogenesis in the
Hypothalamus 55

Endocannabinoids Mediate a Sex Difference in Cell Genesis in the
Developing Amygdala 57

Winged Messengers: Lessons from Birds and Flies 61

Sexual Differentiation of the Neural Circuit for Song in Songbirds 61

Courtship and Copulation in *Drosophila* 62

Sexual Differentiation of the Primate Brain 65

Sexual Differentiation of the Human Brain 69

Androgen Insensitivity Syndrome 72

Estrogen Receptor Mutation and Aromatase Deficiency 73

Congenital Adrenal Hyperplasia 74

Overcoming the Hegemony of Hormones: Genes Matter Too 77

    Epigenetics and the Development of Sex Differences in the Brain 80

The Value of Understanding the Effect of Sex on the Developing Brain 85

References 87

Introduction

Sex—a small word with big meaning. Sex can be a transitive verb, as in to perform the act of sex, or it can be an adjective, as in sex-limited, or it can be a noun, as in the fairer sex. These multiple meanings pervade and often confuse scholarly discussion of the topic, as I experienced recently in a conversation with two colleagues. The first did not know me well and asked, "what do you study?" and before I could answer the second responded with enthusiasm "she studies sex!" Which is true, but not in the way the first colleagues wide eyes and raised brows would imply. I study sex differences, in particular, sex differences in the brain. When I explain this, the response is often, "oh, great, can you explain my husband please?" Which of course, I cannot—nobody can.

While sex differences are self-evident, they have also been largely ignored outside the context of reproductive biology. This began to change in 2001 when the Institute of Medicine published a report *Exploring the Biological Contributions to Human Health: Does Sex Matter?*, in which the overwhelming impact of sex on almost every biological endpoint was examined (Figure 1) (Wizemann and Pardu, 2001). This report also provided the current distinction between *sex* and *gender*. Sex is how we codify the reproductive capacity of individuals based on genetic, gonadal, and somatic characteristics with which we are all familiar. All sexually reproducing species on the planet have two sexes, male and female. There can be some slippage in the constraints of sex, such as sex-changing fish or parthenogenetic female lizards, but there are only two sexes. Gender on the other hand, combines both an individual's perception of self and society's perception and response to that individual as either male or female. Thus, only humans are considered to have gender. But how can we understand gender in all its complexity and conflict in the human animal? The impact of experience, environment, culture, and society on human gender is overwhelming, beginning before birth with the colors chosen for the nursery. It is impossible to parse out the impact of the pervasive and infinitely variable influences that shape the gender of any one individual. But sex is a biological variable with relevance to far more than reproductive capacity and about which appreciation is growing exponentially. Indeed, recent years have seen the development of a scientific society, the Organization for the Study of Sex Differences (http://www.ossdweb.org) and a new journal, *Biology of Sex Differences*. These reflect the increasing awareness of the importance of sex to virtually all aspects of human health, including the brain, be it the developing, mature, or aging brain. But separating

FIGURE 1: Cover of the Institute of Medicine report: "Sex Matters: Assessing the Biological Contributions to Health." This report was commissioned by the Institute of Medicine, an arm of the National Academy of Sciences of the United States and after a multiyear and highly contentious process was released in 2001. The report concluded that the effects of gender or sex were pervasively important to all aspects of human health, and the importance of sex differences in the brain was particularly noted as an area requiring further research. Used with permission of the National Academic Press.

out the confounding influences of being human from the purely biological requires that we turn to animal models.

A wide array of animal models have proven informative about the processes by which males and female are distinguished from each other, ranging from *Drosophila* and *Caenorhabditis elegans* to a variety of rodents and ultimately nonhuman primates, particularly *Rhesus macaques*. The genetic versatility combined with unparalleled fecundity of laboratory rats and mice has naturally led to the predominance of these species for in-depth analyses of the process of sexual differentiation in mammals. The advantages to using rodents are obvious, and while the shortcomings are equally obvious, we set them aside when the goal is to understand brain development at the cellular and molecular level.

Sex Determination versus Sex Differentiation

The production of a male versus a female is the product of two phases, sex determination and sex differentiation. In mammals, sex determination is a function of the sex chromosomes, XY for male and XX for female, a fact startlingly recent in its discovery in the 1950s. Even more startling is that it was not until the 1990s that the gene on the Y chromosome coding for male development was finally identified after an exhaustive search (Sinclair et al., 1990; Classic Reference 1). Termed *Sry*, for sex-determining region of the Y chromosome (Figure 2), this single gene codes for a small transcription factor protein called tdf, or testis-determining factor, which initiates a cascade of gene expression that directs the development of the bipotential gonad to become a testis, as opposed to an ovary, at the beginning of the process of sex differentiation. One of the earliest steps in testis development is the regulation of steroidogenesis, with a down-regulation of estrogen production and increased androgen synthesis. An additional hormone, anti-Müllerian hormone (AMH), produced by the immature testis complements the actions of androgens so that the Wolffian duct system will survive and differentiate into the vas deferens and epididymis of the male, while the female duct system, the Müllerian ducts, will degenerate in response to AMH. Androgens will further the process of differentiation by promoting the formation of a penis and scrotum to form male genitalia, as well as increased bone and muscle mass, and later secondary sex characteristics such as beard growth in men and racks of antlers in deer.

If there is no functional tdf, either due to the absence of *Sry* or a mutation that leads to a dysfunctional protein, the gonadal anlage will develop as an ovary, the Müllerian duct will survive and develop into the oviducts, uterus, and the upper portion of the vagina, the Wolffian duct will degenerate due to lack of androgens, and the genitalia form a clitoris, labia, and the remaining portion of the vagina. Thus, in both males and females, the gonadal, or somatic sex, is secondary to and dissociable from genetic sex. An XX individual with a *Sry* translocated onto an autosome will develop testis, and likewise, an XY individual with a mutated or deleted *Sry* will develop ovaries. So what about the brain?

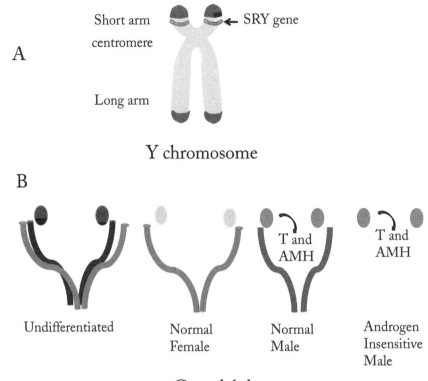

Short arm

centromere

← SRY gene

A

Long arm

Y chromosome

B

T and AMH

T and AMH

Undifferentiated

Normal Female

Normal Male

Androgen Insensitive Male

Gonadal duct systems

FIGURE 2: Sex determination. (A) Sex is determined by the *Sry* gene (for sex determining region of the Y chromosome), which directs the undifferentiated gonad to become a testis. (B) The reproductive ductal systems are present in both males and females initially, with the Wolffian duct system slated to become the epididymis and vas deferens of the male and the Müllerian duct system slated to become the oviduct, uterus, and upper third of the vagina in females. The Wolffian duct is retained in males in response to androgens produced by the testis, and the Müllerian duct degenerates in response to AMH, also produced by the testis. In females, the Wolffian ducts degenerate due to the lack of androgen and the Müllerian ducts are retained due to lack of AMH. If a male has a dysfunctional androgen receptor, the Wolffian ducts degenerate but so do the Müllerian ducts because the testes still produce AMH.

In an interesting analogy to the gonads, the brain is also bipotential, capable of developing a male or female phenotype with equal measure. Rather than being directly linked to the expression of a single gene, however, brain phenotype is largely determined by the hormonal profile generated by the gonads (but see the emerging role of genetics to brain sexual differentiation discussed below). Brain sexual differentiation occurs during a developmental sensitive window, and in another inter-

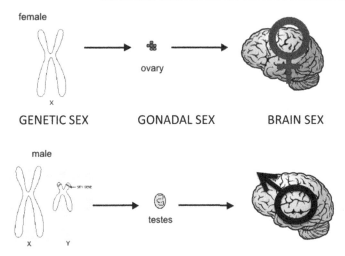

FIGURE 3: Sex differentiation. The chromosomes XX and XY determine the gonads, so that an XY individual (with *Sry*) develops testes while an XX individual develops ovaries. The gonads release hormones very early in development and the hormones then sexually differentiate the brain. In particular, the male testes produce androgens, some of which are converted into estrogens by neurons in the brain, and this masculinizes the rodent brain. The female brain develops largely as the default pathway, although there is increasing evidence for a contribution from ovarian steroids that occurs later in development than masculinization.

esting similarity with the gonads, the female brain develops as the default, that is, in the absence of hormonal changes, whereas the brain is differentiated as masculine by steroid hormones derived from the testis. In this way, a series of sequential events beginning with genetic sex determination at fertilization will progressively differentiate the gonads, the body, and finally the brain into male or female (Figure 3).

Masculinization, Feminization, and Defeminization

Sexual differentiation of the brain can be further subdivided into three distinct processes as a means of providing us a descriptor of the final outcomes (Figure 4). *Masculinization* refers to the retention of the Wolffian duct system and development of the associated secretory glands combined with changes in the brain that will promote the expression of male sexual behavior in adulthood. In our animal models, male sexual behavior is stereotypical and quantifiable, thereby providing a valuable read-out of the process of brain masculinization. *Feminization* is the opposite, the retention of the Müllerian duct system while the Wolffian ducts degenerate, and a neural capacity for expression of female sexual behavior. In rodents, female sexual behavior is a combination of solicitive behaviors, called proceptive behaviors, and lordosis, an arch-backed posture that facilitates mounting by the male and successful copulation. *Defeminization* is the less intuitively obvious process in which the feminine phenotype is actively removed from the male body and brain. The repression of the Müllerian duct system by AMH is an example of defeminization in the periphery and an analogous process occurs in the brain so that the capacity for female sexual behavior is removed. Defeminization occurs because feminization is the default process and must therefore be stopped. A fourth process, demasculizination, is not a naturally occurring component of development because there is no mechanism for stopping an inducible process that does not involve some exogenous intervention. The term *demasculinization* is sometimes used to refer to disruption of masculinization but the utility of this is debatable. In general, masculinization, feminization, and defeminization are best restricted to the discussion of reproductive endpoints and should not be taken as global descriptors of the entire brain, as each process can occur to varying degrees for varying endpoints and varying neural substrates.

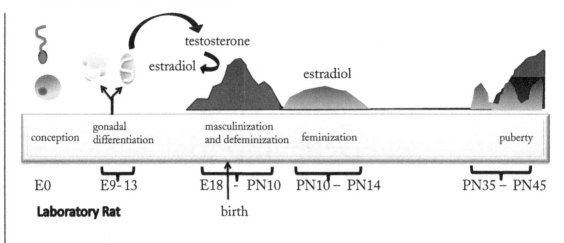

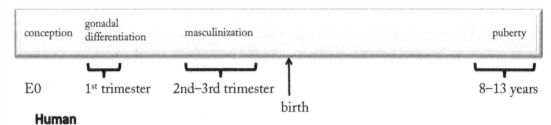

FIGURE 4: Timeline of sexual differentiation in rats and humans. The differentiation of the gonads into either ovary or testis occurs strikingly early in both rodent and primate sexual differentiation. The testes provide the hormones that drive masculinization and defeminization of the brain, which in rodents occurs perinatally, meaning just before and after birth. In humans, all evidence indicates masculinization occurs prenatally. It is not known if there is defeminization in humans. Recent evidence suggests a later period of feminization in the rodent brain that is driven by estrogens, presumably from the ovary. Puberty onset is earlier in females and is marked by the re-emergence of steroid hormone production, which is cyclic in females and tonic in males. In both sexes, adult hormones activate the previously organized neural substrate.

Steroid Hormones are Potent Modulators of Brain Development

Before beginning an in-depth discussion of the mechanisms by which the brain is sexually differentiated, it is essential to have an appreciation for steroid hormones, endogenous signaling molecules that are synthesized in one organ and travel through the bloodstream to ultimately alter the functioning of other organs. There are a number of things that distinguish steroids from other cellular messengers, including that they are synthesized on demand, never stored, are generally long-lived particularly when bound to steroid-binding globulins, and they bind to a family of receptors characterized for their action as nuclear transcription factors. Steroid receptors directly interact with DNA at palindromic sequences located in gene promoters and that are specific to different types of steroid receptors. Thus, the estrogen receptor (ER) binds to one specific sequence of nucleotides, while the androgen receptor (AR) or progesterone receptor (PR) binds to its own distinct sequences. Steroid receptors do not act in isolation but instead have a host of cofactors and corepressors that help determine whether the action on gene expression is facilitatory or inhibitory. Steroid receptors also interact with other well-known components of the transcriptional machinery, such as immediate early genes like *c-fos* or *zenk*, and these associations can determine how the receptor behaves in terms of gene expression. Steroids receptors have revealed themselves to have highly labile morphology, which changes as a function of ligand binding, and this also impacts on their association with other proteins and ultimate responses as agonist or antagonist (Figure 5).

In addition to the varied ways in which steroids and steroid receptors can impact on gene expression, recent years have provided irrefutable evidence that an entirely separate, but no less potent, component of steroid hormone action is initiated at the cell membrane. Both traditional and unique steroid receptors have the capacity to nestle into the cell membrane and, when activated by binding of ligand, will directly associate with various membrane-associated kinases, ion channels, and G-protein-coupled receptors. These effects are considered rapid and can occur within seconds to minutes of steroid binding, which is in direct contrast to the slow and stately progression of changes in gene expression induced by traditional transcription factor actions. Rapid membrane effects are also by their very nature transient, as opposed to gene expression changes that lead to

Two-step model of steroid receptor transactivation in vivo.

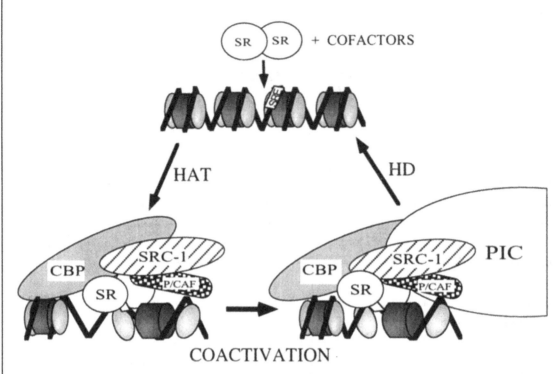

FIGURE 5: Steroid receptors are nuclear transcription factors. Steroid receptors are members of a superfamily of nuclear transcription factors that interact directly with DNA, but this requires that they also associate with other cofactors that possess enzymatic activity capable of modifying the surrounding chromatin. CREB-binding protein (CBP) and steroid receptor coactivator 1 (SRC-1) are two examples of proteins that associate with steroid receptors and allow for transcription by relaxing the chromatin. The chromatin contains histones that can be acetylated by enzymes called histone acetyle transferase (HAT), which open up the DNA for transcription, or deacetylated by histone deacetylases (HD or frequently, HDACs), which tighten up the chromatin and suppress gene transcription (reprinted with permission from Jenster et al., 1997).

new protein synthesis and alter the fate of cells in a myriad of ways. Not all membrane effects are transient, however, as signaling cascades can find their way to the nucleus via a variety of pathways and so-called rapid effects can also have enduring consequences. The current state-of-the-art is trying to discern the relative contribution of the classic genomic actions of gonadal steroids and their receptors versus the newer, more modern view of membrane initiated effects that are more akin to other signaling molecules, including neurotransmitters. One emerging concept is the potential for

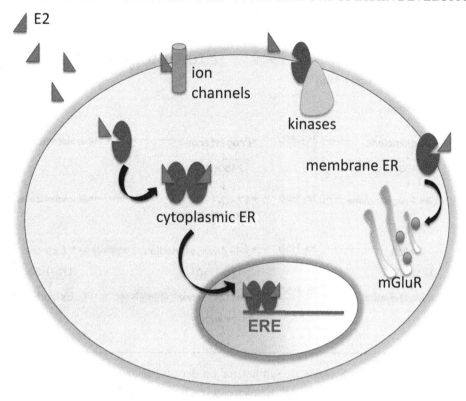

FIGURE 6: Steroid receptors have a multiplicity of actions: The mechanisms of steroid receptor action have expanded beyond being only a transcription factor. We now know steroid receptors can also intercalate into the cell membrane where, after binding steroid, they can directly activate various protein kinases, modulate the activity of ion channels, or interact with specific subunits of postsynaptic neurotransmitter receptors. These actions can both directly alter the excitability of cells as well as activate intracellular signal transduction pathways, some of which might then alter gene expression (reprinted with permission from McCarthy, 2010).

sequential modes of action, with membrane effects serving as a necessary precursor to more enduring classical genomic actions, a scenario that has been demonstrated for some adult hormonally responsive behaviors (Figure 6).

The limiting step in steroid action is availability, and availability is determined by synthesis, transport, uptake, binding, and degradation. Each step is subject to regulation, but our knowledge of that regulation varies considerably. Several of these are essential to the normal progression of hormonally induced sexual differentiation of the brain, including the timing of steroidogenesis, access to the brain from the bloodstream and availability of local converting enzymes and functional

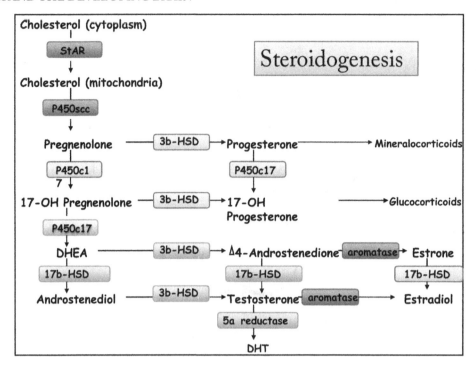

FIGURE 7: Steroidogenesis: Steroids are synthesized on demand following the rate-limiting step of cholesterol transport into the cell by STAR. Cholesterol is a large molecule, and much of steroidogenesis consists of tearing it down step by step, with each removal of a critical carbon or hydroxyl group dramatically altering the bioactivity of the molecule. Once past progesterone, the pathways diverge into those associated with the adrenal steroids, the glucocorticoids and mineralacorticoids, versus those associated with the gonads, androgens and estrogens. All estrogens, including the most potent, estradiol, are derived from androgens and are considered the endpoint of steroidogenesis since further modifications generally render them inactive.

receptors. The expression of essential steroid receptors, such as ER, AR, and PR, are at their highest level in several developing brain regions, as is the critical enzyme, aromatase, which converts androgens into estrogens. Essential steroid-binding globulins uniquely expressed during development serve as a selective filter by sequestering some steroid in the bloodstream while allowing others access to the brain.

The process of steroidogenesis begins with the transport of cholesterol into the cell by a rate-limiting protein called steroid acute regulatory protein (STAR) (Figure 7). Cholesterol is a reasonably large molecule consisting of three aromatic rings and a pentamer atop which sits a long and complex carbon chain. Steroidogenesis consists largely of progressive reductions in that carbon chain and either addition or removal of a few oxygens and hydrogens on the other carbon rings. The initial by-products of cholesterol are the pregnanes, progesterone, and the associated pregnenolones, and from there, the process diverts along two main pathways, one toward the adrenal steroids, i.e., glucocorticoids and mineralocorticoids, and the other toward the sex steroids, i.e., androgens and estrogens. The final end-product is estradiol, the smallest but in many ways most potent of the steroids. From there, additional changes render the steroids largely inactive and increasingly water soluble to promote excretion via the kidneys.

Sex Differences in the Brain Are Established During a Developmental Sensitive Window

We are all familiar with the dramatic events associated with puberty, when the hypothalamic–pituitary–gonadal axis matures and that sweet adorable child turns into a moody unrecognizable teenager enslaved by raging hormones. Most assume this is the first time that hormones have held sway over the tender mercies of their child's brain, but in reality, the reproductive axis is awakening from a long slumber, having been held under active repression since birth. Long before first breath, the impact of hormones have left their mark on the brain. In males, the embryonic testis actively synthesize and secrete testosterone during the latter half of the second trimester in humans and the final days of pregnancy in our animal models, rats, mice, guinea pigs, and hamsters. Levels of circulating testosterone approach that of the adult and persist into postnatal life for a matter of weeks or hours, depending on the species, before being repressed to undetectable levels until the onset of puberty years or months later. The beginning of testosterone production embryonically marks the onset of the sensitive period for brain masculinization, and the closing of the window of opportunity is defined by the point at which exogenous hormone treatment of females is no longer effective at inducing a male-like response (Figure 8).

STEROID LEVELS IN THE DEVELOPING BRAIN

In rodents, the end of the sensitive window varies as a function of endpoint, meaning the sensitive window is not the same for differentiation of male sexual behavior versus control of gonadotropin secretion or growth hormone secretion. How or why this occurs is not intuitively obvious, as one would assume that the sensitive period would be directed by the decline in steroidogenesis by the gonad since the gonad is the sole source of steroid, or is it? We know that developing male brains are exposed to high levels of testosterone from their own testis, with peak levels toward the end of gestation (Weisz and Ward, 1980; Classic Reference 2) and again approximately 2 h after birth (Corbier, 1992; Rhoda et al., 1984), and that circulating testosterone is consistently low in the perinatal female rat (Lieberburg et al., 1979). However, if estradiol is measured in newborn plasma, it

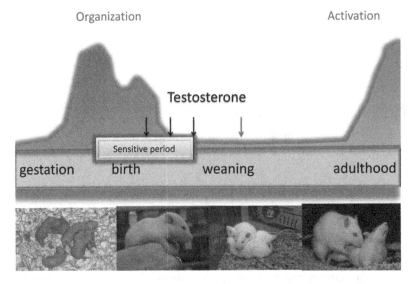

FIGURE 8: The organizational action of steroids occurs during a sensitive period. Just before and after birth, the male testes begin copious production of testosterone, which gains access to the brain and is converted to estradiol and induces masculinization in a process referred to as organization. This only occurs during a sensitive window that is defined by the onset of testosterone production. Treatment of females with testosterone during this period will permanently organize their brains as male (black arrows), but the sensitive period closes when the female is no longer responsive to exogenous hormone injections (red arrows). After this developmental period, steroids drop to undetectable levels until puberty. The increase in steroids at that time will induce activation of the brain to regulate behavior, but the actions of the steroids are constrained by the organizational events that occurred earlier.

is relatively low, but tissue levels in the male hypothalamus are elevated. This can be explained as a product of aromatization in neurons of peripherally derived testosterone (McEwen et al., 1977; Classic Reference 3). Conversion of androgens to estrogens in the brain is common at this age, and the plasma glycoprotein, α-fetoprotein, binds circulating estradiol of maternal or placental origin in order to protect the developing female brain from the masculinizing effects of this steroid (Bakker et al., 2006). Aromatase activity is particularly elevated in neurons of the neonatal preoptic area (POA) and hypothalamus and is detectable in the amygdala, hippocampus, and cortex (Hutchison, 1995; Lephart et al., 1992; Maclusky et al., 1994; Roselli and Resko, 1993). So how long does this period of elevated circulating testosterone in males last?

When peripheral steroid levels are measured in newborn animals, the majority of the data indicates the period of elevated testosterone is exceedingly short, on the level of hours. Whole-body content of androgen (testosterone + dihydrotestosterone) is higher in males on both embryonic days 18 and 19, but is equivalent in the two sexes on later gestational days leading up to birth (Baum et al., 1988). A second period of elevated androgens in males occurs immediately postnatal, although the magnitude of the sex difference is much smaller. Circulating testosterone is elevated in males 2 h after birth (Lieberburg et al., 1979) and whole-body androgen content is higher in males at 1 and 3 h after delivery, but not later (Corbier et al., 1983), while plasma androgen is higher in males only at 1 h postpartum, not 24 h. Thus, the period of sexually dimorphic exposure to elevated circulating androgens appears to be largely prenatal, lasting only up to 3 h after birth by some accounts. A temporary reduction in the capacity of the liver to metabolize steroids to inactive water-soluble by-products is largely responsible for the elevated androgens in the circulation at birth. Within 24 h of birth, the liver resumes metabolism and the sex difference in circulating and whole-body androgen is lost. Moreover, preventing masculinization by removal of the testis is effective only when done immediately after birth, although the impact on different endpoints can vary. The control of gonadotropin secretion and defeminization of sexual behavior are both maximally affected by castration within minutes of birth, with declining effectiveness with as little as a 6-h delay (Handa et al., 1985). Conversely, the volume of the sexually dimorphic nucleus (SDN) in the POA is equally reduced by castration at 0 h as at 24 h, suggesting that variables other than the immediate postnatal rise in testosterone are critical to its differentiation. Consistent with this, inducing full masculinization of the SDN in a female requires steroid treatments beginning prenatally and extending several days postnatally, while the same cannot be said for masculinization of sex behavior. The end of the sensitive period for masculinization of sex behavior, SDN development and gonadotropin secretion are also quite distinct, and recent evidence suggests a later period of feminization is also driven by elevated estrogens (Bakker and Baum, 2008).

The variation in endpoints reveals at least three distinct critical periods for the organizational effects of steroid hormones on sexual differentiation, which correspond to behavioral masculinization, defeminization, and control of gonadotropin secretion. Yet, there has been no empirical explanation as to how a unitary source of steroid, the gonad, could direct these temporally distinct processes. An explanation remains elusive in part because of a lack of information about what the steroid levels are in the actual brain during these developmental time points. A recent exhaustive quantification of multiple brain regions at multiple time points reveals a decidedly different picture of what is occurring centrally versus peripherally. The period of elevated steroid, in particular estradiol, and also testosterone and dihydrotestosterone, is markedly longer than that in the periphery, and the temporal pattern of decline differs across regions (Konkle and McCarthy, 2010). The most parsimonious explanation for the disconnect between central and peripheral steroid levels is independent steroidogenesis within the brain that is not dependent upon peripherally synthesized testosterone. This does not mean it is entirely independent of the gonad, as there may be an important initiating role of peripheral steroid. If the prenatal surge in testosterone is instead differentially regulating local steroidogenesis in the brain, set to begin at discrete times and occurring at different levels, this would provide the desired temporal variability necessary for regional differentiation. Emerging theories on highly localized steroid synthesis in the brain may be fundamental to differences in steroid-induced differentiation in males and females, and this is an important area for future investigation.

ORGANIZATIONAL/ACTIVATIONAL HORMONE EFFECTS ON THE BRAIN

Regardless of where the steroids are ultimately derived from, it is clear that they are only effective at permanently changing the brain during a restricted period of time. The effects of hormones on the brain during this sensitive window are referred to as *organizational*, as these early effects will both limit and direct the effects of steroid hormones on the mature brain, and we refer to those effects as *activational* (Phoenix et al., 1959; Classic Reference 4). Any first foray into the mechanistic basis of a newly observed sex difference in brain, behavior or physiology begins with the question, is it organizational or activational? In other words, some sex differences between adults are the direct result of early hormone exposure, while others are solely the result of the sex differences in adult circulating hormones, i.e., high testosterone in males and fluctuating estradiol and progesterone in females. New investigators to the study of sex differences are encouraged to consult an explanatory guide prepared by leaders in the field (Becker et al., 2005).

Sex Differences in Physiology and Behavior Are Coordinated

The ultimate function of sexual differentiation of the brain is to maximize reproductive fitness, and this is best achieved by assuring that brain sex matches gonadal sex. Successful reproduction requires the production of viable gametes in both sexes, the transfer of those gametes from the male to the female, fertilization, pregnancy, delivery, and in many species, including all mammals, postnatal care. Thus, successful reproduction requires coordination between physiology and behavior to assure that the hormonal events mediating ovulation in females also assure the female is behaviorally receptive to mating with males and, if necessary, will even seek out and solicit matings by males. Once inseminated, the sexual receptivity of the female is shut down in most species, humans being a notable exception, and the hormonal milieu of pregnancy, which is critical to the appropriate development of the fetus and final differentiation of the mammary gland to allow for milk production, takes over. This same hormonal milieu will alter the brain to assure behavior toward the offspring assures their survival. A female mammal that gives birth and then walks away from her newborn cannot be considered to have successfully reproduced, as the offspring will surely die due to hypothermia, starvation or predation. The female must first prepare a safe haven, such as a nest, in anticipation of giving birth; then upon arrival, she needs to clean her newborns, discard or consume the placenta, keep the young gathered in the nest, hover over them to allow for nursing (a posture some call kyphosis); and lastly, she may be called on to defend the nest against aggressors. Indeed, maternal aggression is identified as a specific form of aggression both because of its intensity and its transience; once the young have matured, the female no longer attacks intruders. This rolling program of physiology and behavior reveals the tremendous power of steroids to temporally correlate brain plasticity with specific internal and external demands. However, this plasticity is constrained by sex, or more precisely, by sexual differentiation of the brain. An adult male provided the same hormonal milieu as a sexually receptive or pregnant female will not behave as a female. Moreover, he will not respond physiologically in the same way either, meaning there is no positive feedback of estradiol to induce a luteinizing hormone (LH) surge and hence ovulation and the mammary glands will not readily differentiate and begin milk production. But why? Why is it that males do not respond physiologically or behaviorally to hormones in the same way as females? Granted males do not have ovaries, but ovulation is an end-product of a series of neural event and males have all the neural components. It is the neuronal components that do not respond. A male who has been feminized by being deprived of his testis very early in life will ovulate a transplanted ovary if provided the appropriate hormones. This is because ovulation begins in the brain.

OVULATION BEGINS IN THE BRAIN

The hypothalamic–pituitary–gonadal axis (Figure 9) provides the anatomical locations for the critical components of gamete production. The hypothalamus contains specialized neurons that express gonadotropin-releasing hormone (GnRH), sometimes referred to as luteinizing hormone-releasing hormone (LHRH). The GnRH neurons project to the anterior pituitary and induce release of the gonadotropins into the bloodstream. The gonadotropins are LH and follicle-stimulating hormone (FSH), named for just one of their multiple functions. Once in the bloodstream, LH and FSH travel to the gonads where different cells express G-protein-coupled receptors on their surface for either peptide. In females, the maturing follicle expresses LH receptors, and when it is fully mature and there is a large surge in LH into the bloodstream, the follicle ruptures (due to a chemical reaction called luteinization) and the ova is expelled, i.e., ovulation. If the follicle is not mature or the

---

FIGURE 9: The hypothalamic–pituitary–gonadal axis. Reproduction begins in the brain with specialized neurosecretory neurons that release GnRH into the portal vasculature of the median eminence, a bony structure that provides anchorage for the pituitary at the base of the brain. The anterior lobe of the pituitary consists of highly specialized cells that synthesize large quantities of trophic hormones that, once released into the general circulation, access their target organs and promote growth and other physiological responses. The releasing hormone, GnRH, induces the release of LH, as well as FSH, both of which act on the gonads. The names of the trophic hormones are derived from their action in females. FSH stimulates the growth of ovarian follicles, while LH stimulates the luteinization response on the mature follicle, which is required for ovulation. A large bolus of LH is needed to induce ovulation and is referred to as the LH surge. Lower levels of LH and FSH before the LH surge coordinate to induce steroidogenesis and the production of estrogens and progestins after the LH surge. In males, LH stimulates the production of androgens by the testes, while FSH stimulates spermatogenesis. The stimulation of steroidogenesis is critical to the closing of the loop by exerting negative feedback on the hypothalamus to reduce further gonadotropin production. In males, the hypothalamic–pituitary–gonadal axis is always under negative feedback, but in females, there is a brief period of positive feedback that leads to increased estradiol synthesis and ultimately the LH surge. Positive feedback is a neural phenomenon mediated by the brain. If the capacity for positive feedback is lost as a result of neonatal treatment of females with testosterone or estradiol, she will be sterile despite having normal ovaries due to the inability to ovulate (reprinted with permission from Porterfield, 2003). Used with permission of Elsevier.

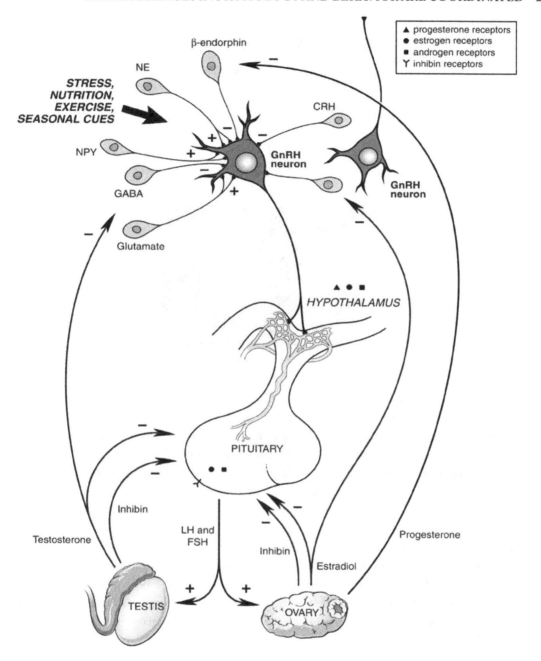

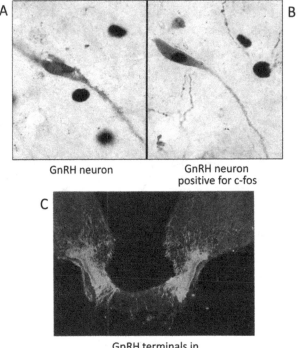

GnRH neuron

GnRH neuron
positive for c-fos

GnRH terminals in
the median eminence

FIGURE 10: GnRH neurons express *c-fos* when activated. The specialized neurons that express GnRH can be visualized in postmortem brain using immunocytochemistry technique, which uses an antibody against the protein of interest. The antibody is attached to a chromogen which can be visualized via a chemical reaction that creates a brown precipitate, in this case, GnRH. If the goal is to visualize two proteins of interest simultaneously, the brown precipitate can be turned black via an additional chemical reaction, in this case, *c-fos*, which is a protein product of an immediate early gene that is expressed when a neuron is activated. The GnRH neurons visualized in panel A are from a female that was not having an LH surge, whereas the neurons visualized in panel B came from a female rat that was euthanized at the time of an LH surge. Antibodies used in immunocytochemistry can also be linked to fluorophores, proteins that fluoresce when excited by the appropriate wavelength of light. In panel C, the terminals of the GnRH neurons (yellow/orange) are visualized as they terminate in the medial portions of the median eminence, an inverse arching structure that provides support for the portal vasculature that transports GnRH to the anterior pituitary (all three photos courtesy of Gloria Hoffman).

LH levels are not sufficiently high, there is no ovulation. The LH surge is directed by the GnRH neurons (Figure 10), which must synchronously fire and release into the anterior pituitary in order to provide a coordinated and massive release of LH into the bloodstream. The organization of the synchronous firing of the GnRH neurons is conducted by estradiol in a process referred to as positive feedback. Most feedback loops involve negative feedback, such as the thermostat in your house

FEMALE REPRODUCTION IS CYCLIC

• positive feedback

• negative feedback

FIGURE 11: Female reproduction is cyclic. Because of the combination of positive and negative feedback effects of estradiol, the female reproductive axis goes through periods of intense activity followed by periods of quiescence that lead to new periods of activity.

which shuts off the heat when the temperature reaches a certain point, thereby maintaining homeostasis. But female reproduction is not about homeostasis (male is though), as it is cyclic, and cycles require a combination of positive and negative feedback (Figure 11). Estradiol provides both positive and negative feedback onto the GnRH neurons in sequential fashion by first coordinating the synchronous firing, while simultaneously building up LH stores in the anterior pituitary. The final trigger for the LH surge is a combination of rising estradiol levels and the circadian rhythm so that in rodents, at least, ovulation can be timed to within a few hours (not true in humans in which the LH surge lasts approximately 24 h). Once ovulation has occurred, estradiol will induce negative feedback, tamping down the firing of the GnRH neuron and allowing the system to reset for new follicular development and a new cycle, provided pregnancy has not occurred.

One of the more surprising aspects of this beautifully coordinated system is that estradiol does not act directly on the GnRH neurons, as they do not express ER. Instead, estradiol acts on surrounding neurons, which then project onto the GnRH neurons and thereby control their activity. There are multiple classes of neurons on which estradiol appears to exert an influence that is critical to the functioning of GnRH neurons, in both males and females, but the control of the LH surge involves a selective group of neurons located in a specific subnucleus called the anteroventral periventricular (AVPV) nucleus. Neurons here project directly to the GnRH neurons and regulate their activity. The sexual differentiation of the AVPV is complex and multifactorial, but a consensus is emerging that this may be the basis for the difference in the positive feedback effects of estradiol in female but not male brains.

Once ovulation occurs, the ruptured follicle takes on a new life and transforms into a unique endocrine organ called the corpus luteum, for "yellow body," which describes its appearance as a lumpy lipid-rich protuberance on the ovary. Its appearance belies its function as a veritable steroid-producing machine (recall steroids are closely related to lipids, being derived from cholesterol). High levels of progesterone will prepare the uterus in the event of fertilization and subsequent implantation. During this time, both estradiol and progesterone are exerting negative feedback effects on the GnRH neurons, as there is no need for another LH surge. If fertilization and implantation occur, the membranes associated with the implanting fetus and the developing placenta take over from the corpus luteum and provide the hormones needed to maintain the pregnancy for the duration. If fertilization and implantation do not occur, the corpus luteum collapses and the stage is set for the return of positive feedback effects of estradiol and a new cycle. Hope springs eternal, or at least until menopause.

FEMALE SEX BEHAVIOR IS COORDINATED WITH OVULATION

While this hormonal dance is deftly coordinating the physiology of reproduction, so is it coordinating the necessary behaviors. Female rodents, and most mammals with the exception of humans for that matter, are sexually receptive only when there is the potential for pregnancy, meaning just before, during, and after ovulation. Female rodents will solicit the attention of males with a series of proceptive behaviors, which, in rats, consist of locomotor patterns that display rapid hops and darts combined with freezing. Other rodents, such as mice and hamsters, use other stereotypical behaviors and exude powerful pheromones that signal to males their readiness to mate. In some mammals, the simple act of not moving away from a male when he approaches is sufficient to signal a female's willingness to engage. These behavioral components leading up to mating can be classified as appetitive behaviors, contrasted to consummatory, which is the actual act of mating. In female rodents, this involves the assumption of a specific posture called lordosis, a curving of the back and elevation of the haunches to allow the male access. The frequency and intensity of the lordosis posture in females provides a reliable readout of sexual receptivity and is quantified as a quotient as well as a rating (Figure 12). The expression of both proceptive and receptive behavior by females requires a closely controlled temporal exposure to sequential estradiol followed by progesterone. In rats, 48–72 h of estradiol exposure followed by 4 h of progesterone will produce a maximum lordosis response. Physiologically, this precisely mimics the periovulatory period, as it mimics the buildup of estradiol during the positive feedback phase and the increased progesterone production associated with ovulation and formation of the corpus luteum. Outside of this temporal window of hormonal exposure, the female will respond to the male with kicks, boxing, rolling on her back, or just plain running away if given the opportunity. Quantifying lordosis behavior in adults provides an easily attainable and accurate measure of the degree of brain feminization during development. A demonstration of behavioral testing for lordosis behavior is seen in the following video (Video 1).

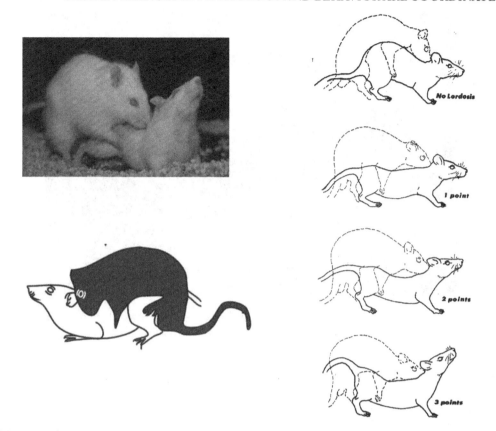

FIGURE 12: Rat sex behavior. The level of rat sex behavior is usually tested by placing a stimulus animal in a glass aquaria, introducing the test animal, and monitoring their interactions. When the male is the test subject, the stimulus female is hormonally primed to be sexually receptive. When the female is the test subject, the stimulus male has had extensive previous experience. The motor components of male rat sex behavior involve mounting the female, intromitting (insertion of the penis), and ejaculation, followed by a postejaculatory interval. The number of mounts, intromissions, and ejaculations during a defined period of time is taken as a measure of the males' behavior. The latency to the first expression of each and the duration of the postejaculatory interval are additional quantitative components of a male's performance. Female sexual behavior involves a proceptive component in which she solicits the interest of the male by hopping and darting and wiggling her ears. Receptivity is expressed by adopting the lordosis posture, an arching of the back, elevation of the rump, and deflection of the tail. During a testing period, the male is allowed to mount the female a specified number of times, usually 10, and the number of times the female responds with lordosis is taken as a quotient expressed as follows (lordosis quotient [LQ]= number of lordoses/10 mounts × 100). Thus, an LQ of 100 indicates the female is highly sexually receptive, an LQ of 50 means she is only moderately receptive, and an LQ = 0 indicates a complete lack of receptivity. The degree to which the back is curved and the rump elevated is scored on a scale of 0 (*no curvature*) to 3 (*most extreme curvature*), and this too can be used as a measure of receptivity and is expressed as the lordosis amplitude (reprinted with permission from Hardy and DeBold, 1973). Used with permission of Elsevier (Hardy and DeBold, 1973).

VIDEO 1: Video of male and female rat normal sex behavior. http://tinyurl.com/mccarthy-video1.

MALE PHYSIOLOGY AND BEHAVIOR ARE NOT TEMPORALLY CONSTRAINED

The hormonal control of male reproduction is far simpler. The GnRH neurons, which show no overt differences from those in females, regulate regular pulsatile release of LH and FSH from the anterior pituitary, and via the bloodstream, these gonadotropins reach the testis and stimulate both sperm production and testosterone synthesis. The feedback effects of testosterone and its metabolite, estradiol, are exclusively negative, maintaining a steady state of gamete and hormone production. Likewise, males steadily and reliably engage in sexual behavior, which can also be broken down into appetitive and consummatory components. In many animals, males must display or show courtship behaviors to impress prospective females. In rodents, males pursue females and engage in anogenital investigation. Males also exude pheromones, which are attractive to females in a reproductive ready state. The consummatory components are broken down into mounting, intromitting (insertion of the penis), and ejaculation. The frequency of each and the latency to first occurrence upon exposure to a female are all used as quantitative measures of male sexual behavior. Higher numbers of mounts with a shorter first latency indicate stronger masculinization of the neural substrates mediating sexual behavior.

One interesting and as yet unexplained difference in the hormonal control of male versus female sexual behavior is the degree of constraint imposed by hormones. As we discussed, females only exhibit receptivity during a narrow window closely tied to ovulation. Moreover, the first time females exhibit the behavior is pretty much as good as the hundredth time; in other words, there is no real learning involved and lordosis is often referred to as a reflex. By contrast, learning plays a very important role in males, with experience producing linear improvement. But even more interestingly, while testosterone exposure, for a period of weeks, is required for the initiation of male sexual behavior, once established in an individual, it will continue for weeks or even months following castration. This is true in rodents as well as primates, including humans. In other words there is no ongoing need for steroids to maintain the behavior, and it only gradually wanes, in stark contrast to the active inhibition of sexual behavior in females that occurs approximately 24 h after the period of receptivity. This marked difference in the hormonal dependence of male versus female sexual behavior presumably reflects some fundamental difference in the neural underpinnings that are sexually differentiated during development, but that difference is remains elusive.

CHANGES IN THE BRAIN INDUCED BY STEROIDS DURING DEVELOPMENT DIRECT ADULT PHYSIOLOGY AND BEHAVIOR

The earliest bona fide neuroanatomical sex differences were reported in the 1960s and were small in magnitude and subtle in nature, having been detected by the high-resolution technique of electron microscopy (EM). During the subsequent 10 years, it became apparent there was a "not seeing the forest for the trees" scenario in that many of the most robust brain sex differences were best observed at a distance. Observing that male song birds sing a complicated repertoire of songs, while females do not, Fernando Nottebohm and his graduate student at the time, Arthur Arnold, at Rockefeller University in New York, reasoned that there must be a neural underpinning to this difference, and they were right. Nissl staining to visualize the gross architecture of the brain revealed several nuclei that are markedly larger in males than females (Nottebohm and Arnold, 1976; Classic Reference 5). Since that time, an entire song control circuit has been well characterized in birds and has provided many important insights into hormonal effects on brain development. Moreover, the report by Nottebohm and Arnold prompted researchers interested in sex differences in mammalian brain to take a step back and look again. This is precisely what Rodger Gorski did, a young investigator at UCLA, leading him to discover what he termed the *sexually dimorphic nucleus of the preoptic area*, or SDN for short (Gorski et al., 1980; Classic Reference 6). This collection of neurons is three to seven times larger in males than females and, despite its small size and ambiguous function, has sparked a cottage industry of study that has established many of the basic principles of sex differentiation. Before delving into greater detail into the specifics of steroid-mediated sexual differentiation of the brain, it is worth a brief detour into the world of transgenic mice and what they have and have not taught us to date.

Knockouts of the Rule: Mice with Null Mutations of Steroid Receptors, Steroidogenic Enzymes, and Binding Proteins

The study of genetically modified mice with null mutations of specific genes has created paradigm shifts in many biomedical disciplines, and this is no less true in the study of sexual differentiation of the brain. Study of some knockout mice confirmed the preexisting dogma. For instance, loss of the α isoform of the ER, referred to as ERKOs (ERα knockout mice), results in males with greatly reduced sex behavior, although they retain simple mounting behavior (Ogawa, 1998; Ogawa et al., 1997). Knockout of the β isoform, creating mice called BERKOs, has no impact on male sex behavior but when combined with ERKOs for a double knockout, meaning both ER isoforms are dysfunctional, total loss of male sex behavior results (Ogawa et al., 2000). A role for ERβ that is distinct from ERα is the induction of defeminization of sex behavior (Kudwa et al., 2005). The importance of estradiol, as opposed to ER, to masculinization of sex behavior is confirmed by the disruption of the gene coding for aromatase (Bakker et al., 2003), the enzyme required for estradiol synthesis, and a long suspected role for some level of estradiol in female brain development is confirmed as well. Disrupting the AR, whether through naturally occurring mutation or by design, also predictably impairs male sexual behavior (Juntti et al., 2008), but conclusions are confounded by the inability to distinguish between developmental organizational and adult activational effects, a caveat that currently applies to all studies of genetically modified mice but promises to be resolved in the near future with advances in temporally and spatially restricted gene inactivation (Juntti et al., 2010).

All mammals use olfaction for intraspecies communication and monitoring the local environment, but mice particularly excel at the detection of odors, which they achieve via a main and accessory olfactory system. The study of olfactory communication in mice has a long and rich history that reveals novelties such as pregnancy block, estrus suppression, estrus induction and synchronization, and gender recognition (for a review, see Brennan and Keverne, 2004). Distinguishing what is mediated by the nasal epithelium and the main olfactory bulb versus the vomeronasal organ and the accessory olfactory system is an ongoing effort. The transient receptor potential 2C (TRP2C) channel is highly localized to the dendrites of vomeronasal sensory neurons, and when rendered ineffective, male mice become notably unaggressive and mount males and females with equal vigor (Leypold et al., 2002). The bisexual behavior of the TRP2C knockout mice does not appear to be due to the inability to distinguish males from females, given that the complete removal of the vomeronasal organ does not impair this capacity across a wide range of species, including mice. Instead, it is the males' preference for females that is lost (Pankevich et al., 2004). How the vomeronasal organ, in particular, the TRP2C channel, regulates sexual and aggressive motivation remains unknown. Further roiling the field are observations of female mice lacking TRP2C function exhibiting high levels of male-like sexual behavior (Kimchi et al., 2007). The authors suggest the vomeronasal organ acts a repressor of male sexual behavior in females and that by removing this inhibition a latent circuitry controlling masculine responses is released and a parallel circuit controlling feminine responses (maternal behavior and nesting) is repressed. On the face of it, this is a reasonable conclusion based on the data and long-held assumptions about how the brain regulates sex behavior. Indeed, the notion that distinct male- and female-specific circuits, analogous to the Müllerian and Wolffian duct systems of the reproductive tract, has its origins in the earliest studies of hormonally mediated sexual differentiation and observations of females exhibiting male-like mounting have pervaded the literature for 50 years. However, to date, the discussion has occurred in the absence of any empirical evidence that there are in fact two separate neural circuitries controlling sex behavior. There are numerous robust and reliable sex differences in brain regions critical for the expression of sex behavior, as reviewed here, and this is often equated with the concept of sex-specific circuits because it is so intuitively appealing. But consider the nature of the differences. As already noted, one of the most celebrated is the SDN-POA because it is three to five times larger in males, but females have an SDN too, it is just smaller. Likewise, males have two to three times more dendritic spines on the POA and ventromedial nucleus (VMN) neurons, but again, females have plenty of dendritic spines and attendant synapses, just not has many as males. Thus, instead of two distinct neural circuits, it is equally likely that there is only one neural network, and that it is differentially weighted toward sex-specific responses as a function of early organization, adult context, and hormonal activation. This idea is not new (Simerly, 2002), but it is not well integrated into the current mindset on sexual differentiation of behavior. The conceptual disconnect between what is known about the neuroanatomy and what is perceived to regulate hormonally mediated sex differences in behavior highlights how far we have to go in developing a unified theory of sexual differentiation of the brain.

Steroids Influence Multiple Endpoints to Organize the Brain

The study of sex differences draws heavily on the advances of developmental neuroscience, which provides an overall understanding of the genetic and environmental causal variables that organize the progression from a simple neural tube to a fully developed nervous system. Embedded within that process is the differentiation of the brain phenotype as male or female. Our global view of sexual differentiation of the brain has now advanced sufficiently to allow for generalities about which developmental processes are critical and which are only modestly involved. Most notable is the importance of hormonal modulation of naturally occurring cell death to regulate size of particular regions and the role of cell-to-cell communication to determine synaptic patterns that are distinctly different between males and females. Also important are sex differences in the relative frequency of distinct neurochemical phenotypes of neurons and their projections, resulting in neuronal networks controlling complex behaviors.

STEROIDS ORGANIZE THE DEVELOPING BRAIN BY ALTERING CELL SURVIVAL

Estimates are that 50% of the cells generated in the nervous system are ultimately fated to die during early periods of naturally occurring cell death, and in several brain regions, the sex of the animal is a major variable determining that fate. Analyses of various systems, including birds and mammals, reveal that volumetric sex differences are established when males and females begin with the same number of neurons but that differential hormonal exposure results in sex differences in cell death. The identification of apoptosis as a mechanism for establishing sex differences in the vertebrate nervous system was made over 20 years ago (Nordeen et al., 1985; Classic Reference 7) but has been investigated mechanistically only recently. The spinal nucleus of the bulbocavernosus (SNB) consists of motor neurons that innervate the penis. The size of the SNB is substantially larger in males, which is not surprising. Ciliary neurotropic factor (CNTF) is up-regulated by androgens in the bulbocavernosus muscle, which is innervated by the SNB motor neurons and then retrogradely transported to act on the CNTF receptors on the motoneurons, promoting their survival. Mutant mice lacking receptors for CNTF have no sex difference in the size of the SNB. CNTF administered to females rescues the motoneurons, and treating males with antagonists to CNTF receptor reduces the number of motoneurons (Forger, 2006; Forger et al., 1995). Thus, testosterone regulates cell death in a specific CNS region because it has evolved control of a well-established neurotrophic mechanism that controls apoptosis in numerous neural tissues.

Developmentally restricted sex differences in cell death leading to adult volumetric dimorphisms have been well characterized and well reviewed for several other brain regions, most notably, the SNB and SDN-POA of rats, the AVPV of mice and rats, and the song control nuclei of birds (Davis et al., 1996; Forger, 2006). In each case, there is a temporal disconnect of days between hormone exposure and maximum cell death, suggesting an initiation of a long-term cellular process by estradiol or testosterone, perhaps involving afferent input or efferent connections. The delay in dying further complicates attempts at identifying the mechanism of hormone action, which can vary from neuroprotection (SDN, SNB, song nuclei) to death promoting (AVPV). Mice with a null mutation in the *Bcl-2* gene, a potent inhibitor of cell death, or in *Bax*, a promoter of cell death, have been usefully exploited to advance our understanding (Forger et al., 2004). Sex differences in the SNB and AVPV were eliminated in *Bax−/−* mice, indicating *Bax* is required for sexually dimorphic cell death in the mouse forebrain and spinal cord. Interestingly, *Bax* is involved in death that is increased by estradiol (in the AVPV) as well as that decreased by testosterone (in the SNB). One advantage of this approach is that the number of neurons observed in *Bax−/−* adults represents the original number generated in each sex, whereas the difference in cell number between *Bax−/−* and *Bax+/+* adults reveals the total number of neurons lost "integrated over the entire developmental cell death period" (Forger, 2006), further supporting the notion that sex differences in cell death contribute to volume differences in multiple brain regions (Figure 13). However, it would be simplistic to suggest

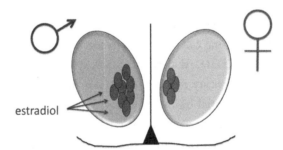

FIGURE 13: Estradiol promotes both cell survival and cell death. The SDN of male rats is much larger than the females because estradiol promotes the survival of these neurons, and more survive in males than females, resulting in a larger overall volume for this structure. The opposite situation occurs in the AVPV, where estradiol actively induces cell death in specific subclasses of neurons, resulting in a smaller AVPV in males compared with females.

that this is the only basis for volumetric sex differences. Brain nuclei are defined by a combination of their histological appearance, the neurochemical phenotype of the cells and the function attributed to that group of cells. They are not monomorphic collections of identical neuronal phenotypes. The ability of steroids also to distinguish the phenotype of neurons, and perhaps even to determine the fate of precursor cells as neurons versus glia, is an important and still relatively poorly understood mechanism for distinguishing male from female brains.

STEROIDS ORGANIZE THE BRAIN BY ALTERING CELL PROLIFERATION

The relative timing and rate of the birth of new neurons is fundamental to growth of specific brain regions. Alterations in the timing and frequency of cell birth and the length of the cell cycle can have profound impacts on the structure and function of specific brain regions. Among the most robust sex differences in the brain, and often those that attract the most interest, are those in which a structure is larger in one sex than the other, most commonly, the male. There are three essential ways a structure could be larger in one sex versus the other: (1) more cells are generated in one sex, (2) more cells differentiate into the dimorphic cell type and/or migrate to a particular region in one sex, or (3) greater cell atrophy or death occurs in one sex. There is strong consensus across brain regions and species that the differential hormonally mediated cell death is the primary determinant of volumetric sex differences. Confidence in this conclusion includes direct observation of little to no difference in detection of new cells but large sex differences in the number of dying cells or loss of cells in numerous systems (for reviews, see Forger, 2006; Simerly, 2002; Morris et al., 2004). Nonetheless, sex differences in cell proliferation may contribute to the modest sex differences in the hippocampus, where male rats have a 10%–15% larger volume (Isgor and Sengelaub, 1998). Within 24 h of birth, male rats show more newly divided cells, and treatment of females with testosterone immediately after birth increases the number of newly born cells to the same level as males. Quantification of pyknotic, or dying cells, at the same time indicates no sex difference or hormonal modulation (Zhang et al., 2008), although a potential role for differential survival mediated by steroids cannot be entirely ruled out. More recently, we have determined an unambiguous role for estradiol in higher rates of cell proliferation and ultimately more neurons born in neonatal males (Figure 14). But there are several aspects of this finding that remain puzzling, including the observation that males make close to twice as many new neurons as females over an extensive period of time, presumably this would result in a hippocampus twice as large. But this is clearly not the case, the male hippocampus is only slightly larger than that in females, and that is on a good day. So what are all these new neurons doing in males and are there other neurons dying off that we have not detected? Most neurogenesis beyond that associated with embryonic brain development has been associated with some form of olfactory learning or with emotional affect. It is possible neonates are learning odors,

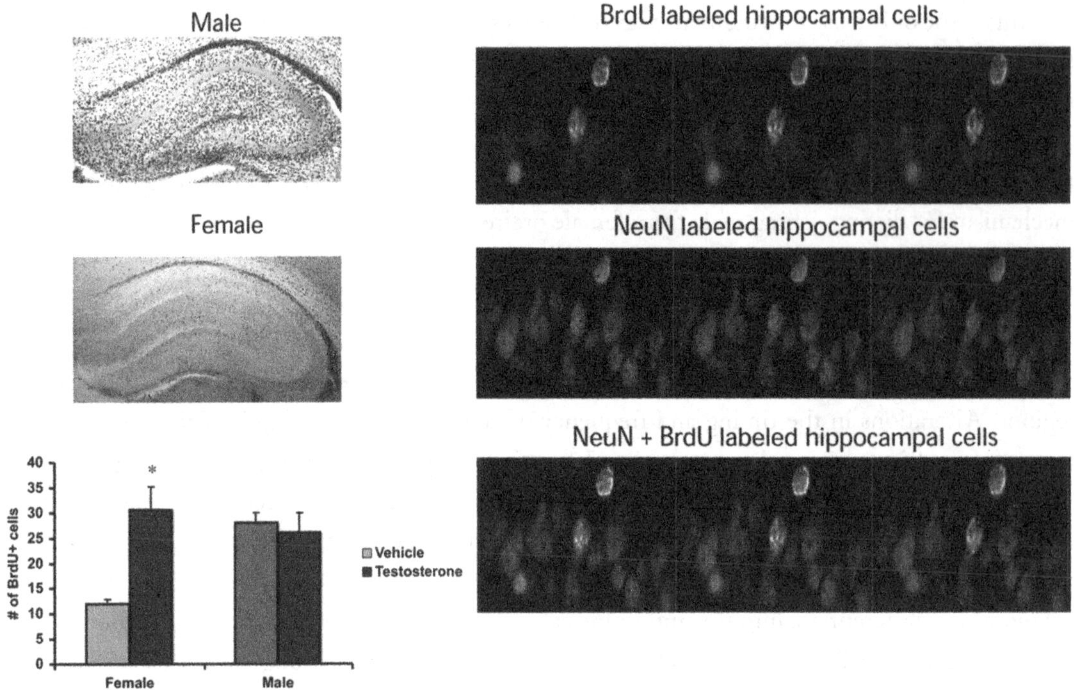

FIGURE 14: Estradiol promotes cell proliferation. The birth of new cells can be quantified by injecting animals with an analog to the DNA nucleotide uracil, called bromodeoxyuracil, or BrdU. When the cell divides and the DNA replicates, the BrdU is incorporated as part of the new DNA and can be visualized later as a brown precipitate in the nucleus of the cell, which here appears as small brown dots in the coronal sections through the hippocampus (upper left). By counting the number of cells with a brown nucleus, the number of new cells born at a given snapshot in time can be determined. Comparison of the number of new cells in the hippocampus of newborn males and females reveals that males make more new cells than females, and treating females with testosterone, which is aromatized to estradiol, increases the rate of cell proliferation to that seen in males. To determine if these new cells will become neurons, a waiting period of at least 2 weeks is required, but fortunately, the BrdU persists. In the images on the right, the BrdU is labeled with a green fluorescing fluorophore, while the neuron specific marker NeuN is labeled with a protein that fluoresces red. When the two are combined, the product is yellow light, as the image in the bottom panel shows at least two of the new cells born in this animal on postnatal day 1 have become neurons 3 weeks later. The images to the right were generated on a confocal laser scanning microscope and can be read left to right as three separate scans moving through the z plane of the tissue, which is required to determine that two fluorophores are genuinely colocalized as opposed to immediately adjacent or on top of each other.

such as that of the dam, and it is also possible that the olfactory imprinting is different for male pups than females, but making these types of causal connections awaits future research.

NEURONAL MIGRATION IS NOT STRONGLY REGULATED BY STEROIDS

Establishing that either hormones or genetic sex influence the migratory pathway of neurons has proved technically challenging because it is difficult to recognize specific types of cells during their differentiation and migration. Lack of strong evidence in favor of clear sex differences in migratory rates or pathways led to a de-emphasis of this as an important variable (Burek et al., 1997), but not because there was strong evidence against it. The advent of live cell fluorescent imaging and transgenic mice has allowed for a reexamination of cell migration and revealed that estradiol decreases the average movement of neurons in the POA, which may contribute to the greater cell density seen in males (Knoll et al., 2007). Based on the embryonic time points examined, which included periods before the onset of sex differences in gonadal steroid synthesis, the authors also propose that the hormonal effects observed are superimposed on a preexisting genetic bias. Thus, this study is one of the first to reflect the new synthesis of direct genetic and hormonal factors in the study of the causes of sex differences in the brain.

STEROIDS REGULATE TROPHIC FACTORS AND ACTIVITY-DEPENDENT SURVIVAL

One of the major variables determining if neurons will live or die is the level of exposure to trophic factors. The best evidence for a hormonally mediated effects on growth factors relevant to the establishment of sex differences is in the spinal nucleus, the SNB (reviewed above), and in the song control nuclei of the zebra finch (Contreras and Wade, 1999; Dittrich et al., 1999; Fusani et al., 2003). Estradiol also increases the amount of brain-derived nerve growth factor in the developing hippocampus (Solum and Handa, 2002) and midbrain (Ivanova et al., 2001) of rodents, but these have not been directly tied to sex differences. Whether estradiol is directly inducing the transcription of growth factors or indirectly having an impact by altering the excitability of a cell and/or its synaptic connections has not been well explored.

Steroids potently modify neuronal excitability in the adult brain, but establishing the same principle in the developing brain is more challenging. The amino acid transmitter, gamma-aminobutyric acid (GABA), is the primary mediator of excitation in the immature brain before the formation of sophisticated synaptic networks, and its excitatory action is enhanced in both duration and magnitude by estradiol (Nunez and McCarthy, 2009; Nunez et al., 2005; Perrot-Sinal, 2001). The depolarizing effects of GABA in the developing brain are tightly linked to establishment of synaptic patterning and considered to serve a vital trophic function (Akerman and Cline, 2007). Thus, the

GABA system is perfectly situated to be an important conduit for hormonally mediated sexual differentiation, and it plays a critical role in the establishment of sex-specific synaptic density in the arcuate nucleus (Mong et al., 2002). Potential additional roles are inferred from sex differences in the response to GABA agonist administration and in GABA levels (McCarthy et al., 2002b).

Once considered the primary source of inhibition in the brain, we now know GABA to be a principal source of excitation via depolarization-induced calcium influx through voltage-sensitive

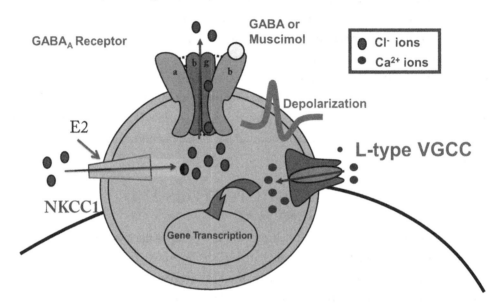

Excitatory GABA Mediates Ca²⁺ Influx

FIGURE 15: GABA is excitatory in the developing brain. The GABA$_A$ receptor forms a channel that is freely permeable to chloride, which flows down its concentration gradient. In mature neurons, the extracellular chloride is higher than intracellular, and so, chloride flows in and hyperpolarizes the membrane, making it more difficult to achieve an action potential. In immature neurons, a transporter called NKCC1 actively moves chloride into the cell so that it builds up a transmembrane concentration gradient that will promote efflux of chloride upon opening of the GABA$_A$ receptor channel. The movement of negative ions out of the cell depolarizes the membrane sufficiently to open voltage-gated calcium channels, resulting in influx of calcium, which initiates various signal transduction cascades associated with growth and differentiation. On occasion, the depolarization induced by the efflux of chloride is sufficient to generate an action potential in the immature neuron. Estradiol further enhances depolarizing GABA action by increasing the activity of the chloride co-transporter NKCC1, thereby further differentiating the transmembrane chloride concentration gradient, which increases the magnitude of the depolarization and the amount of calcium that enters the cell.

calcium channels. This action is most prominent developmentally and appears to be present throughout the brain. The calcium influx induced by the depolarizing GABA action contributes to the maturation of synapses (Ganguly et al., 2001). The excitatory effects of GABA are mediated by the $GABA_A$ receptor, a chloride ionophore, and the relative transmembrane chloride gradient. Whether $GABA_A$ receptor activation results in chloride influx or efflux is determined by the trans-membrane chloride concentration gradient, which is, in turn, determined by the activity of chloride cotransporters (Ganguly et al., 2001; Plotkin et al., 1997a,b; Rivera et al., 1999). During the neo-natal period, the reversal potential for chloride (E_{Cl^-}) is positive relative to the resting membrane potential (Barna et al., 2001), resulting in a net outward driving force upon chloride when $GABA_A$ receptors open and membrane depolarization sufficient to open voltage-sensitive calcium channels, primarily of the L-type (Ikeda et al., 1997; Leinekugel et al., 1995; Obrietan and van den Pol, 1995; Owens et al., 1996) (Figure 15). As development progresses, E_{Cl^-} becomes negative relative to the resting membrane potential, thus shifting the driving force on chloride to inward and leading to $GABA_A$ receptor-mediated hyperpolarization. On the day of birth, pyramidal neurons of CA1 hip-pocampus are synaptically silent but engage in paracrine communication via nonvesicular GABA release (Demarque et al., 2002). The chronic release of GABA, presumably from interneurons, and stimulation of neighboring pyramidal neurons is essential to dendritic arborization and synaptic specialization. How this communication is modulated to provide phenotypic variability has not been addressed. Our observations of a potent enhancement of depolarizing GABA action by es-tradiol and a sex difference in GABA content in the developing hippocampus (Davis et al., 1999) suggests this steroid could be a major contributor to normal hippocampal development.

STEROIDS IMPACT ON AXONAL PROJECTIONS, DENDRITIC BRANCHING, AND CONNECTIONS

In addition to steroid-induced regulation of neurotrophins that regulate cell survival, steroids appear to alter trophic factors that control axonal outgrowth. The principal nucleus of the bed nucleus of the stria terminalis (pBNST) projects to the AVPV as part of a neural circuit controlling gonadotro-phin secretion from the anterior pituitary. The AVPV is a critical node for the induction of the surge in LH release, which is required for ovulation. AVPV neurons are largely glutamatergic and project to the vicinity of the LHRH neurons, which, in turn, project to the anterior pituitary and regulate LH release. There is no compelling evidence for sex differences in the LHRH neurons themselves. However, one of the most robust morphological and functionally significant sex differences in the brain is the 10-fold larger pBNST to AVPV projection in the male. Clever use of explant cultures in which male and female pBNST and AVPV could be mixed and matched, revealed that estradiol was acting in the AVPV to produce a signal to attract the growing axons of the pBNST neurons (Ibanez et al., 2001). The identity of the diffusible factor remains to be determined (Figure 16).

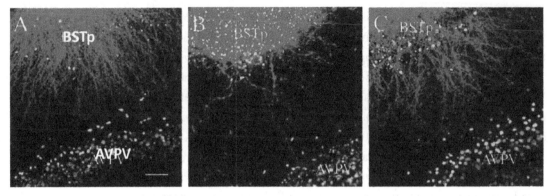

FIGURE 16: Estradiol induces a target-derived diffusible factor that promotes innervation. The principal subdivision of the bed nucleus of the stria terminalis (BSTp) is located dorsal to the preoptic area and contains GABAergic neurons that project to the AVPV nucleus, a critical nucleus in the control of the LH surge. The size of the GABAergic projection from the BNST to the AVPV is almost 10 times larger in males than in females, and this is due to a diffusible factor synthesized and released by the AVPV neurons that attracts the BNST neurons to them and promotes innervation. The synthesis of the diffusible factor is promoted by estradiol. That this was indeed a target derived factor was demonstrated with the use of explant cultures as shown here. The BSTp is labeled in red, while the AVPV is in green. The left image is from a male, the middle is from a female (no growth), and the right is from a female treated with male hormones (photo courtesy of Rich Simerly, University of Southern California).

On a more local level, a sex difference in the number of branches off the primary dendrites of hypothalamic neurons in the rat appears to be partly determined by an estradiol-induced down-regulation of two proteins, paxillin and focal adhesion kinase (FAK) (Speert et al., 2007). Initially identified for their role in cancer metastasis, both FAK and paxillin interact with the integrins and have been implicated in axonal growth and dendritic branching (Rico et al., 2004). Both proteins must be down-regulated for growing processes to detach from the substrate in order to extend or branch (Webb et al., 2004). In an unusual example of hormone action in the developing brain, estradiol down-regulates both these proteins, and females show a developmental peak of expression around one week after birth that seems to be actively suppressed in males by the actions of estradiol (Speert et al., 2007).

Estradiol and insulin have long been known to have a synergistic effect on neurite growth in fetal hippocampal or cortical explants, an effect now known to be the result of an interaction between insulin-like growth factor (IGF-1) receptors and ERs, presumably at the membrane (reviewed by Toran-Allerand, 2005). These two receptors appear to act in tandem to promote cell survival and neurite outgrowth in a variety of brain regions, with considerable emphasis placed on a potential neuroprotective effect in the adult (Cardona-Gomez et al., 2000, 2002). An effect of estradiol on neurite growth and innervation is also found in the songbird system, with the added twist that the estradiol appears to be derived centrally within the brain rather than from testosterone from the testes (Holloway and Clayton, 2001). Indeed, the notion that local de novo steroidogenesis occurs in the developing songbird brain is becoming increasingly accepted (London et al., 2006) and emerging as a potential convergence point in the effects of direct genetic and gonadal hormonal factors leading to sex differences in the brain (Wade and Arnold, 2004). A major challenge ahead is identifying the factors that control steroidogenesis in the brain. Equally important is determining whether such steroids contribute to establishing sex differences as opposed to having equal actions in males and females or alternatively act in a sex-specific manner to make males and females more similar than they otherwise would be (McCarthy and Konkle, 2005).

STEROIDS ORGANIZE THE DEVELOPING BRAIN BY ALTERING SYNAPTIC CONNECTIVITY

Equally important as the number of neurons in a particular brain region is the number and type of synaptic connections they receive and project to other areas. The nature of synapses is determined by two variables, the neurotransmitter utilized and the geographic location on the cell being synapsed upon. Classic neurotransmitters are biogenic amines, serotonin, dopamine, noradrenaline, and acetycholine. Peptides are also neurotransmitters, including oxytocin, vasopressin, and releasing hormones such as corticotrophin-releasing hormone, GnRH, and thyroid-releasing hormone. These neurotransmitters are synthesized in specific cells and have well-defined neuroprojections that deliver the transmitter to areas where the cognate receptors are found on the postsynaptic cell. Both classic and neuropeptide transmitters are critical components of specific physiological and cognitive processes. In parallel to these specific modulators are the ubiquitous amino acid transmitters, GABA and glutamate, which are generally inhibitory and excitatory. Some estimate that virtually every synapse in the brain contains at least one of these amino acid transmitters, which can be released in combination with other transmitters and modulate the postsynaptic response. As a result, synapses are generally inhibitory or excitatory, meaning the stimulus that triggered vesicular fusion and neurotransmitter release is either blocked or dampened in the postsynaptic cell or it is passed along or even amplified.

The number, strength, and flavor of synapses are directly related to changes in physiology and behavior and thus of are of great interest empirically. But quantifying synapses presents numerous technical challenges and each approach has inherent limitations and rests on underlying assumptions. A tradeoff exists between characterizing the neurochemical profile and efficacy of a small number of synapses versus quantifying large numbers of ill-defined synapses that represent the population (Figure 17). Useful information has been gained using both approaches in the developing brain but the characterization of large numbers of synapses has is more suited to understanding the establishment of sex differences in synaptic profiles.

Synapses are predominantly formed on the dendritic tree but also occur on the soma and at the axon hillock, a point close to the soma that is particularly effective for inhibition. Within the dendritic tree, there is further division, with synapses occurring either on specialized protuberances called dendritic spines or on the intervening shafts (Figure 18). The synapses on dendritic spines are generally excitatory and use glutamate as neurotransmitter, whereas those on the soma are generally inhibitory and use GABA as neurotransmitter. Neither distinction is absolute, however, with spinous synapses found on the soma and GABAergic synapses present throughout the dendritic tree. Regardless, it is valuable to quantify synapses in the context of where they are localized on the postsynaptic cell, and this can be achieved with the labor intensive but precise approach of EM.

Sex differences in the brain were first reported in the 1960s by Donald Pfaff (1966; Classic Reference 8) and followed shortly thereafter (5 years was considered short back then) by Raisman

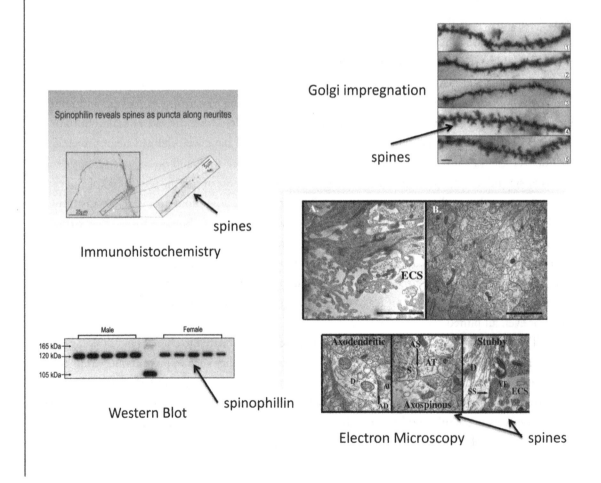

and Fields (1971; Classic Reference 9), in which they examined synaptic profiles based on EM observations. While these generated great interest, they were, as noted above, not of a very impressive magnitude or tied to any physiological function. In the 1980s, a series of exhaustive studies conducted by a Japanese group led by Yasumasa Arai were published to much less fanfare but clearly demonstrated remarkable degrees of divergence in the synaptic profiles in brain regions directly associated with reproduction. In the arcuate nucleus, a brain region critical to the control of the anterior pituitary, Arai and his colleague Akira Matsumoto painstakingly determined that males had almost twice as many axosomatic (on the cell body) synapses as females, whereas females had almost twice as many dendritic spine synapses compared with males (Matsumoto and Arai, 1980; Classic Reference 10). There was no sex difference in the number of dendritic shaft synapses, and the total number of synapses was the same in males and females. This latter point is particularly important, as it demonstrates that only the type, not the amount of synaptic activity, varies between the sexes. Moreover, Matsumoto and Arai further demonstrated that the hormonal milieu in the first days of life determined the adult synaptic pattern so that females treated with testosterone as neonates had a male-like pattern, whereas males castrated as newborns had a female-like pattern. This was the

---

FIGURE 17: Quantification of dendritic spine synapses. Quantifying the number and/or density of dendritic spine synapses can be achieved in multiple ways. The most accurate but also most difficult is the use of EM to visualize individual synapses and determine their functionality by the presence of synaptic vesicles and a postsynaptic density. The disadvantage of this approach is the inability to quantify large numbers of synapses over a broad area in multiple individuals. Golgi impregnation provides visualization of the entire neuron and under sufficient magnification can be used to count the number of spiny protuberances from a specified length of dendrite. This allows for determining both the total number of dendritic spines and the number per unit length of dendrite, i.e., density. The image here is from a Purkinje neuron in the cerebellum. The ability to also quantify dendritic length, thickness, and branching in the same cell is an added benefit. This same approach can be used in neurons filled with dye from an electrode after recording. Western blot is a technique for quantifying the amount of protein in a tissue sample by binding it with an antibody and visualizing a chromophore associated with the initial (primary) antibody. Spinophillin is a protein that is preferentially associated with dendritic spines and can be used as a proxy marker for the number of spines in a given sample. The advantages of this approach are its ease of use in a large number of samples. The disadvantage is the lack of any information about the identity or morphology of the neurons in the sample. Lastly, immunocytochemistry for spinophillin on cultured neurons can be used to visualize spin-like processes that can then be counted. The advantage of this approach is that neurons can be treated in a dish, avoiding the influence of other brain regions or signals from the periphery, but this can also be a disadvantage as it is an artificial situation and should therefore be complimented by other approaches (photos courtesy of Stuart Amateau and Jessica Mong).

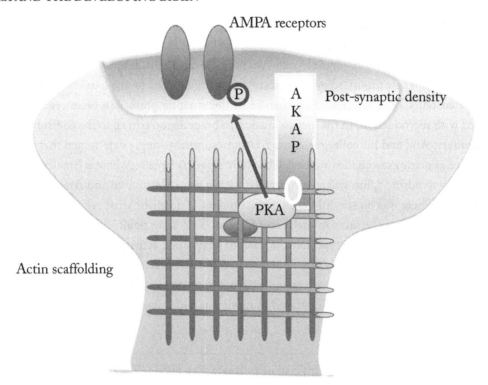

FIGURE 18: Dendritic spine synapses. Synapses that occur at the end of small (<5 μm) protuberances from dendrites are of particular interest because they have been shown to form and disappear rapidly and in response to external and internal stimuli, such as hormones, fear, stress, learning, parenting, etc. They are also of interest because they tend to be excitatory, meaning glutamate is the dominant neurotransmitter, and because they are neatly compartmentalized, which makes them easier to study. Glutamate receptors of the AMPA type are anchored in the postsynaptic density by an actin filament network that is maintained by phosphorylation from the kinase PKA. Stimuli that change the activity of PKA can thereby alter the kinetics of glutamate receptors in the synapse, which, in turn, alters both its functionality and its durability.

first demonstration that the adult synaptic profile was established neonatally and is also probably the first demonstration of the principle of developmental organizational effects of steroids determining the neural substrate upon which adult steroids activate behavior. Matsumoto and Arai went on to provide similarly exacting data on sex differences in synaptic profiles in several other nuclei critically involved in reproductive physiology and behavior.

Understanding the rules of synapse formation remains a fundamental goal of neuroscience. Early notions that growing axons plugged into dendritic trees to "hard wire" the brain into a permanently established circuit have been largely supplanted by the widely held view that synapses are "plastic" or even transient. The discovery of the dynamic synaptic profile of hippocampal pyramidal neurons in reproductively cycling female rats was an important contributor to our new view of the plastic adult brain (Woolley, 2007). But the hormonal modulation of synaptogenesis in the developing brain appears to be an entirely different circumstance because the patterns established will endure for a lifetime. As noted above, landmark studies in the early 1970s and 1980s by the teams of Raisman and Field (1971) and Matsumoto and Arai (1980) described marked hormone-determined sex differences in the synaptic profile of specific brain regions. The mechanisms orchestrating organization of synaptic profiles in neonates appear to be distinct from those in the adult because some steroid-modulated cell signaling in neonates has not been found in adults to either increase or decrease synapses.

STEROIDS ORGANIZE THE DEVELOPING BRAIN BY ALTERING NEUROCHEMICAL PHENOTYPE

An early focus in studies of sexual differentiation was sex differences in neurotransmitter systems, which were nearly all found to be either sexually dimorphic or modulated by steroids in some brain region. These include serotonin, noradrenaline, dopamine, *acetylcholine, GABA, oxytocin, vasopressin, calcitonin gene-related peptide,* galanin, met-enkephalin, β-endorphin, neuropeptide Y (De Vries et al., 1984), and most recently, kisspeptin (Kauffman et al., 2007). Sex differences are found in the level of the ligand, the receptor binding profile, the amount and activity of the synthetic enzymes, and in reuptake and degradation. These findings speak to the pervasiveness of steroid hormone action throughout the brain, and the multiplicity of levels on which they act. Despite all this attention, we know relatively little about how steroids organize sex differences in neurotransmitter systems and even less about how developmental steroid exposure determines sex differences in adult neurochemical phenotype.

The neurotransmitter phenotype of a particular neuron provides its individual identity, defining and limiting its role in the brain. Neurons with the same neurochemical phenotype tend to live together in discrete neighborhoods, coalescing to form a specific nucleus and generate projections to other distant neighborhoods made of other groups of like-minded neurons. The size of a neighborhood is largely determined by the number of neurons of a particular phenotype, and this can be determined in two ways: (1) the number of cells that are born and/or die when the nucleus is being formed, as discussed above, or (2) the number of cells that differentiate into that phenotype during the maturation process. Evidence for the first strategy is self-evident in the role of cell death in determining the size of the SDN-POA, which consists entirely of GABAergic neurons, and in the AVPV, which is a complex mixture of dopaminergic, GABAergic, and kisspeptin-expressing neurons. The neurochemical profile of the AVPV is not entirely explained by differential cell death, however, and is providing the means for attacking the much more challenging question of whether steroids influence phenotypic differentiation of particular neurons.

One of the more profound neurochemical sex differences in the brain consists of vasopressinergic innervation of the lateral septum. In a wide range of species including reptiles, amphibians, numerous bird species, and the majority of mammals examined to date, males have more vasopressin (or vasotocin in nonmammals)-producing cells, which project more densely to critical brain regions (reviewed by De Vries and Panzica, 2006). In mammals, these cells are found in the BNST and amygdala and densely project to the lateral septum (Figure 19). The vasopressinergic system illustrates numerous general principles of sexual differentiation but also provides multiple examples

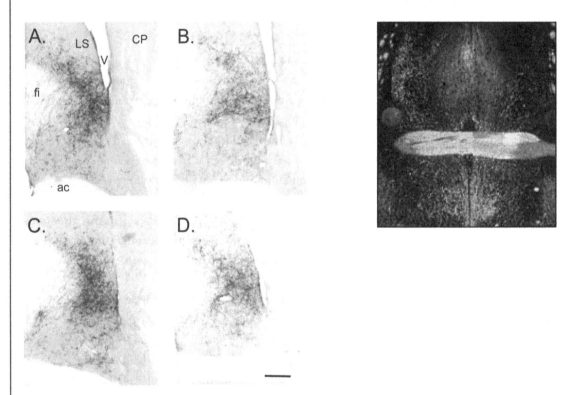

FIGURE 19: Vasopressin. The neuropeptide vasopressin is synthesized by neurons in the brain that project to the posterior pituitary and release into the blood stream, but these same neurons also project throughout the brain and the pattern of the projections is different in males and females across a wide range of species. The establishment of the sex difference in vasopressin innervation is the result of a complex interplay of genetics and hormones during development and in adulthood (reprinted with permission from de Vries et al., 2008). Copyright 2008, The Endocrine Society.

of exceptions or unique components of the process that remind us not to overgeneralize the origins and maintenance of sex differences in the brain.

Vasopressin Is a Model of Steroid-Mediated Sexual Differentiation of the Brain

- Males have more vasopressin expressing neurons than females.
- Vasopressin's cell number and innervation are organized by gonadal steroids during a developmental sensitive window.
- Sex differences in adult vasopressin innervation are activated by circulating gonadal steroids.

Vasopressin Demonstrates Unique Parameters Associated with Sexual Differentiation of the Brain

- XX versus XY genotype contributes to the sex difference in mammals.
- There is no role for differential cell birth or cell death.
- Steroids differentiate neurons to be vasopressinergic but also impact expression in adulthood.
- The purpose of the sex difference is to entice males to behavior more like females.

Each of the unique aspects requires considerable discussion and will be revisited as we approach those topics, but for the current topic, the vasopressin system is an excellent example of steroids impacting the neurochemical phenotype of neurons, thereby generating an important functional sex difference.

The kisspeptin system may provide another example of steroid-mediated neurochemical differentiation, but the jury is still out on this conclusion. There is a large sex difference in the number of kisspeptin neurons found in the AVPV, with females having more than males (Figure 20). As noted above, the AVPV is unusual in that its overall size is larger in females than males, and this appears to be a function of greater cell death in males due to early hormone exposure. But the period of cell death occurs before the onset of kisspeptin expression, which is closer to puberty than the sensitive period for sexual differentiation. Whether neurons that are ultimately destined to express kisspeptin selectively die off in males or whether steroids act to program kisspeptin expression at a future date remains unknown at this time.

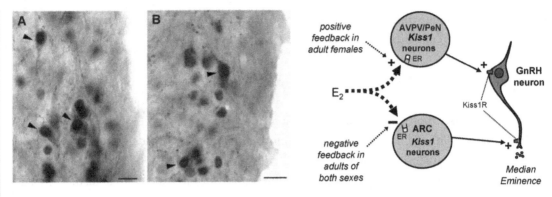

FIGURE 20: Kisspeptin. The neuropeptide kisspeptin is critically important to puberty and the coordinated control of gonadotropin secretion in males and females. There are substantially more neurons that express kisspeptin in the AVPV of females than males, and this may be the neuroanatomical basis of the sex difference in positive versus negative feedback effects of estradiol. In the upper left panels, kisspeptin is visualized by immunocyochemistry and appears as a brown precipitate in individual neurons. The black reaction product is evidence of ER on the left and PR on the right, which are localized on the nucleus of the kisspeptin neurons, demonstrating the potential for direct steroid regulation (reprinted with permission from Clarkson et al., 2008). The schematic on the right illustrates the central role of kisspeptin (Kiss1) in the control of the GnRH neurons that project to the anterior pituitary via the median eminence (reprinted with permission from Kauffman, 2010). Used with permission of the Society for Neuroscience (Clarkson et al., 2008); Used with permission of Elsevier (Kauffman, 2010).

Cellular Mechanisms of Steroid-Mediated Organization of the Brain

The previous section has discussed in considerable detail the myriad ways that steroids exert an influence on the developing brain. In summary, steroids induce sex differences in the brain by modifying all of the predicted parameters: cell birth, cell death, cell differentiation, axonal growth, dendritic branching, and synaptogenesis (Figure 21). Even glia do not escape the pervasive and enduring hegemony of hormones during the perinatal sensitive window for sexual differentiation of the brain. Once sex differences are well characterized, the next most critical question is, what are the steroids really doing to bring about these changes? In other words, what is the mechanism? Some might denigrate the importance of understanding mechanism, asking, what difference does it make to determine how a steroid is acting as long as we know it is? The answer is that we cannot truly understand why until we know how; moreover, we will never gain a genuine understanding of the relative contributions of nature versus nurture if we do not know all the variability inherent in nature. The consequence of a single unified mechanism of steroid action throughout multiple brain regions to impact multiple endpoints is markedly different from regionally specific mechanisms that invoke multiple distinct cellular pathways that provide numerous nodal points for interjection of genetic or experiential influences. We are sufficiently far along in our understanding at this point to come down firmly on the side of multiple distinct cellular pathways, but we still have a long way to go before we rest. With each new brain region or endpoint that is explored, surprisingly novel and unexpected mechanisms are emerging and are telling us that the variables regulating sex differences in the brain are not set apart from those regulating normal brain development but are instead uniquely intertwined and yet distinct from those pathways mediating normal brain development.

PROSTAGLANDINS MASCULINIZE THE PREOPTIC AREA AND SEXUAL BEHAVIOR

One of the more surprising findings of cellular mediators of sexual differentiation of synaptic patterning and adult behavior was the discovery that prostaglandin E2 (PGE2) is the masculinizing agent of the rat POA, downstream of estradiol that derives from testicular testosterone (Amateau and McCarthy, 2004). Prostaglandins are derived from arachidonic acid in the lipid membrane via

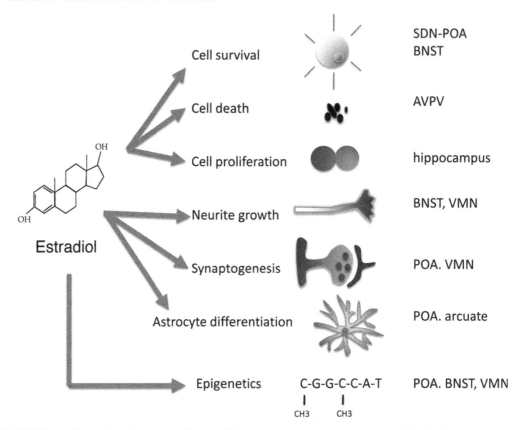

FIGURE 21: Steroids influence multiple cellular endpoints to organize the brain. As the principal mediator of sex differentiation of the rodent brain, the majority of research to date has focused on the cellular mechanisms of estradiol action. Paradoxically, estradiol both stimulates and promotes cell death at the same time, albeit in different subregions of the POA. Recent evidence reveals that estradiol also promotes cell proliferation in yet another brain region, the hippocampus. In another paradox, estradiol increases dendritic spine synapse formation in the POA and VMN of the hypothalamus, while suppressing the formation of dendritic spines in the arcuate nucleus. This may, in part, be a function of estradiol's effects on astrocytes, which is to increase the degree of complexity by promoting growth and branching of primary processes. Lastly, the enduring effects of estradiol on neuronal morphology may be mediated by epigenetic changes to the DNA and surrounding histones.

the cyclooxygenases, COX-1 and COX-2. The latter is considered an immediate early gene responsive to a variety of stimuli including infection, injury, and stimuli associated with neuronal plasticity (Hoffmann, 2000). As early as the day of birth, COX-2 mRNA and protein are significantly higher in the POA of males, and treating females with estradiol increases COX levels to that of males within 48 h. Increased COX-2 is directly correlated with increased PGE2 production, and treating newborn females with estradiol increases PGE2 levels almost sevenfold. Most importantly, administrating PGE2 to newborn females induces a permanent twofold to threefold increase in dendritic spines on neurons in the POA, but not the VMN or the hippocampus. Females treated with PGE2 as neonates will then express robust male sexual behavior as adults if provided the proper hormonal milieu and salient cues. Conversely, blocking PGE2 synthesis temporarily in newborn males by inhibiting the COX enzymes significantly reduces POA dendritic spines to the level seen in normal females and severely impairs the expression of male sexual behavior in adulthood (Amateau and McCarthy, 2004; Todd et al., 2005).

The induction of dendritic spine synapse formation by PGE2 involves glutamate signaling, specifically the α-amino-3-hydroxyl-5-methyl-4-isoxazole-propionate (AMPA) receptor subtype, and interestingly, not the *N*-methyl-D-aspartate (NMDA) receptor, which together with AMPA receptors is essential to formation of dendritic spines in the VMN (Todd et al., 2007). The receptor to which PGE2 is binding has also been identified and is a combination of the EP2 and EP4 receptors. Prostaglandin receptors are notoriously promiscuous and engage in a great deal of crosstalk so it is impossible in this case to genuinely distinguish the need for one over the other. But it really does not matter since they both link to adenylyl cyclase and promote the production of cAMP. Of course, cAMP is well-known for its ability to activate the kinase, protein kinase A (PKA), and here is where we come full circle. Some forms of PKA specifically aggregate in dendritic spines where they associate with the cell matrix and promote the insertion of AMPA receptors into the postsynaptic membrane. If PKA is removed from the dendritic spine or otherwise inhibited, the AMPA receptors internalize and in essence fall out of the synapse, rendering it mute to glutamate (Burks et al., 2007; Wright and McCarthy, 2009). Conversely, when PKA is activated, for example, in response to PGE2-induced activation of EP2 and EP4 receptors, AMPA receptors are inserted into the synapse, and these both promotes the formation and stabilization of dendritic spines. This is believed to be the mechanism by which estradiol promotes increased numbers of dendritic spine synapses in the developing POA (Figure 22). An animated review of this sequence of events can be found in Video 2.

The real proof that this scenario is in fact the cellular cascade the mediates masculinization of behavior comes in measuring the adult behavior of animals manipulated as neonates to either stimulate this pathways in the absence of steroid or block the pathway selectively. This has been achieved at multiple nodal points on the pathway, providing visually compelling evidence of the importance of prostaglandins in the developing brain to adult sexual behavior (Video 3).

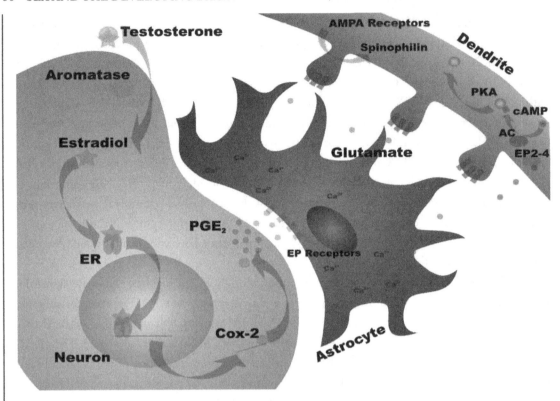

FIGURE 22: Prostaglandins mediated estradiol-induced synaptogenesis in the developing preoptic area. During neonatal rodent development, a principal action of estradiol after it is aromatized from testosterone in the POA is the induction of the enzyme COX-2, a critical enzyme in prostaglandin synthesis. The prostaglandin PGE2 binds to EP2 and EP4 receptors, which are linked to adenyl cyclase, cAMP production, and kinase PKA activation, which anchors AMPA receptors in the postsynaptic membrane via phosphorylation. PGE2 also acts on astrocytes and induces glutamate release. Together, these coordinated events promote the formation and stabilization of dendritic spine synapses on neurons of the POA, and this increase in synapses is functionally correlated with increased male sexual behavior in the adult (reprinted with permission from McCarthy 2010).

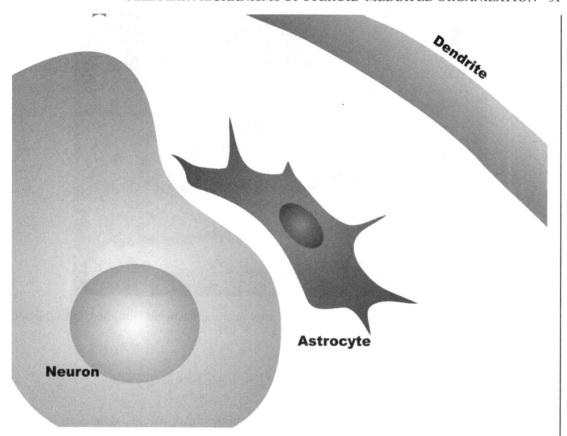

VIDEO 2: Cellular cascade in the POA. http://tinyurl.com/mccarthy-video2.

VIDEO 3: Absence of adult male sexual behavior after neonatal treatment with the COX inhibitor, indomethacin. http://tinyurl.com/mccarthy-video3.

GABA INDUCES SEX DIFFERENCES IN ASTROCYTES IN THE ARCUATE NUCLEUS

Although much attention has focused on sex differences in the shape, size, and number of neurons, there are equally robust morphological differences in the astrocytes in several brain regions, including the POA, arcuate nucleus, and hippocampus (reviewed by McCarthy et al., 2002a). It is not clear whether the changing shape of astrocytes causes or is caused by changes in neurons, but the communication between the two cell types appears essential for sex differentiation in some brain regions. This is perhaps most clearly established in the arcuate nucleus, where a neuronal factor, GABA, is up-regulated in response to estradiol, is released from neurons, and then acts on astrocytes to increase the number of processes and frequency of branching (Mong et al., 2002). Prostaglandins are also emerging as important mediators of neuronal/glial cross-talk and this is a likely component of the synaptogenic effect observed in males. Morphological characterization in the newborn rat POA reveals a greater arborization of male astrocytes. Treating females with masculinizing hormones mimics the morphology of males, and giving females PGE2 also induces an astrocytic phenotype closer to (but not identical) that of males. Sex differences in glia appear to precede sex differences in neurons in at least one nucleus in songbirds (Nordeen and Nordeen, 1996), suggesting in some instances glia may be the primary target of steroid-induced differentiation. The detection of ERs and the appearance of aromatase enzyme in astrocytes following traumatic injury (Garcia-Segura et al., 1994; Jordan, 1999) has further increased attention on this previously neglected cell type.

In many instances, it is unclear whether steroids are acting directly on astrocytes or indirectly via messages sent from neighboring neurons. This is not the case in the arcuate nucleus where it is clear that estradiol acts first in neurons to promote the synthesis of GABA by increasing the amount of the rate limiting enzyme, glutamic acid decarboxylase. This enzyme is strictly neuronal, so there is no question that the primary site of action is neurons. But astrocytes have receptors for GABA, specifically GABA$_A$ receptors, and when activated, these receptors initiate a cellular cascade that

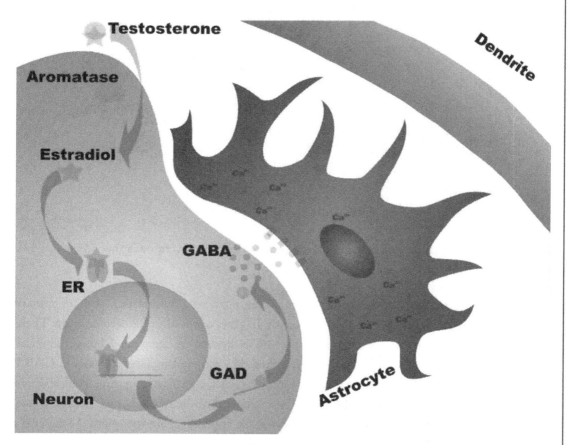

FIGURE 23: GABA mediates estradiol-induced astrocyte differentiation in the developing arcuate nucleus. During neonatal rodent development, a principal action of estradiol after it is aromatized from testosterone in the arcuate is the induction of the enzyme glutamic acid decarboxylase (GAD), which is the rate-limiting enzyme in the synthesis of the amino acid transmitter, GABA. The enzyme GAD is found exclusively in neurons, but GABA acts on neighboring astrocytes to induce the growth and branching of primary processes, resulting in a stellate appearance. The increased growth of astrocytes is associated with a reduction in the formation of dendritic spine synapses in this brain region, which is closely associated with the anterior pituitary and the release of trophic factors (reprinted with permission from McCarthy, 2010).

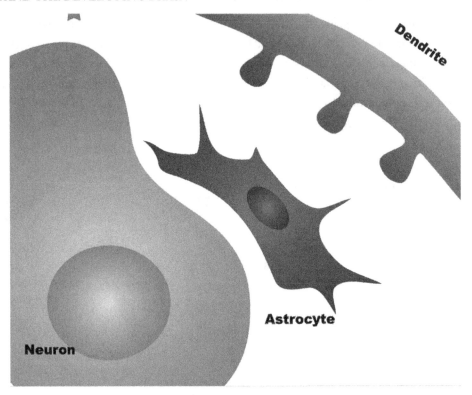

VIDEO 4: GABA-induced differentiation of astrocytes. http://tinyurl.com/mccarthy-video4.

leads to increased stellation of the astrocytes. Stellation refers to a combination of more primary processes that are longer and branch more frequently. In other words, the cells become bushy. There is a marked sex difference in how bushy astrocytes are in the arcuate, being much bushier in males than females (Mong et al., 1996), and this is determined by estradiol-induced increases in GABA (Mong et al., 2002). How or if the change in morphology of the astrocytes, in turn, mediates a sex difference in the synaptic profile of the arcuate is unknown, but the two measures are tightly and inversely correlated, the more complex the astrocytes, the fewer the dendritic spines (Mong et al., 1999). Either way, this is a clear example of the requirement for cell-to-cell communication for sexual differentiation of at least one cell type in the brain, astrocytes (Figure 23, Video 4).

GLUTAMATE RELEASE IS CRITICAL TO SEX DIFFERENCES IN SYNAPTOGENESIS IN THE HYPOTHALAMUS

The VMN is located in the mediobasal hypothalamus and considered the principal brain region controlling female sexual behavior (Mathews and Edwards, 1977; Pfaff and Sakuma, 1979a, b) and is therefore central to feminization. Whether the neurological underpinnings of defeminization reside in the same brain regions controlling female sexual behavior is unknown, but it is a logical place to start investigating. The neurons of the VMN are sexually dimorphic, with males having dendrites that branch more frequently and therefore have more overall dendritic spine synapses than females (Mong et al., 2001; Schwarz and McCarthy, 2008; Todd et al., 2007). This sex difference is established by testosterone aromatized to estradiol during the perinatal critical period in the same manner as in the POA. Importantly, there is no role for PGE2 in the establishment of this sex difference. Thus, one hormone, estradiol, induces both masculinization and defeminization, but the cellular mechanism(s) mediating these two processes are distinct. Estradiol-mediated defeminization begins with a rapid (~1-h) ER-mediated activation of PI3 kinase and enhanced release of presynaptic glutamate. The connection between PI3 kinase activation and glutamate release remains poorly characterized but is independent of protein synthesis. Increased synaptic glutamate leads to increased activation of postsynaptic NMDA receptors followed by dendritic branching and construction and stabilization of dendritic spines (Schwarz et al., 2008). The latter process is dependent upon protein synthesis in the postsynaptic neuron but does not require ER in the postsynaptic neuron (Figure 24). Activity-dependent dendritic growth and synaptogenesis is a common theme throughout brain development, but in the case of sexual differentiation, it is quite surprising that initiation begins with nongenomic effects of estradiol. The effects of estradiol on the presynaptic neuron are receptor mediated as demonstrated by the ability of the selective ER antagonist ICI 182,780 to block estradiol-induced glutamate release. There are two major ER isoforms, ERα and ERβ. ERα is the predominant isoform in the developing hypothalamus (Shughrue et al., 1997) and activation of ERα using the selective agonist, PPT, mimicked the effects of estradiol (Schwarz et al., 2008). Thus, while the effects of estradiol are nontraditional, the receptor mediating the effects is the classic ER. These findings are also important in demonstrating that estradiol-mediated sexual differentiation of the brain is not a cell-autonomous process in which only ER-containing neurons change morphology in response to steroid exposure. Instead, a neurotransmitter serves as a signaling factor that elicits a morphological change in an entire network of cells. This results in an entire population of differentiated neurons as opposed to the differentiation of a distinct neuronal circuit embedded within a population of undifferentiated neurons (Video 5). This has important implications for how the brain regulates sexual behaviors by suggesting that all neuronal inputs to sexually differentiated brain regions such as the POA and VMN are considered and interpreted depending on the animal's sex, regardless of whether the incoming signal are relevant to sex. Both the POA and VMN subserve multiple other functions, including maternal behavior, temperature regulation, and feeding, to name a few. Among the major challenges ahead is determining how inputs as various as olfaction and fear are processed by neurons in a sexually dimorphic way to produce the appropriate behavioral response.

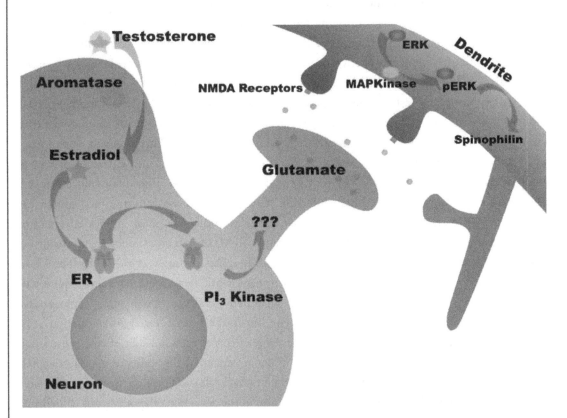

FIGURE 24: Estradiol promotes glutamate release from developing hypothalamic neurons. In this brain region, estradiol has an entirely different type of action after it is aromatized from testosterone. Instead of inducing gene transcription, the activated ER interacts with PI3 kinase, which, via mechanisms that remain poorly understood, induces glutamate release from the presynaptic terminals that binds to AMPA and NMDA glutamate receptors on the postsynaptic side and promotes the formation and stabilization of dendritic spine synapses. Concurrent with the increased synaptogenesis is dendritic growth and branching, resulting in overall more synapses but at the same density. In other words, males have more dendritic spine synapses because they have more dendritic length.

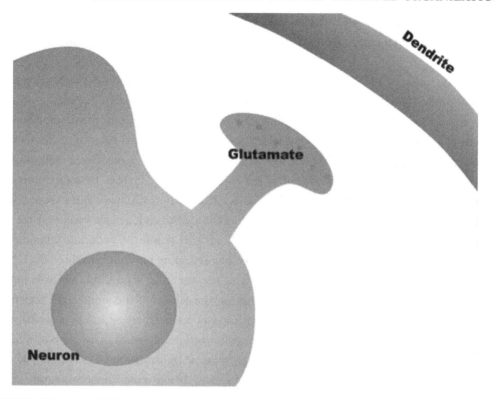

VIDEO 5: Glutamate differentiation of neurons. http://tinyurl.com/mccarthy-video5.

ENDOCANNABINOIDS MEDIATE A SEX DIFFERENCE IN CELL GENESIS IN THE DEVELOPING AMYGDALA

The amygdala is a brain region noted for its role in emotional responding, in particular, both the response to and encoding of fear. Things that are scary are very potent stimulators of learning, meaning this is something to be avoided in the future, and the amygdala is central to that type of learning—often called fear conditioning. The amygdala is also critical to social behaviors such as play by juveniles and sex between adults. The olfactory stimuli that is so critical to arousing the interest of a male mouse requires processing by the amygdala before it works its magic. Given how important the amygdala is to things that differ in males and females, it is not surprising that it is also resident to some degree of sex differences, but the differences are relatively small and undramatic. A great deal, and perhaps inappropriate, amount of attention is given to any sex differences that involve a brain region being larger in one sex versus the other and the amygdala gained some fame for its ability to change size in adulthood in response to changes in steroid levels (Breedlove, 2001; Cooke et al., 1999). The amygdala is also notable for displaying a detectable degree of lateralization, meaning there are differences between the left and right hemisphere amygdala and the magnitude of sex differences in synaptic profile differs between the two sides as well (Cooke et al., 2007).

The discovery of sex differences in the rate of new cell proliferation in the developing hippocampus prompted a similar analysis in the amygdala. While this is still an emerging story, it is proving to be a uniquely fascinating one. There are more new cells born in the developing medial amygdala of female versus male rat pups, which makes this an unusual reversal of the traditional sex bias right from the start. But there are two additional aspects that make it even more surprising. First is that the sex difference is mediated by the endocannabinoid system. Endocannabinoids are lipid membrane-derived signaling molecules that act at ultrashort distances and have vanishingly short half-lives. As the name implies, endocannabinoids are the endogenous variant of the mind-altering cannabis drugs found in marijuana. There are two dominant endocannabinoids, anandamide and 2-acylglycerol (2-AG). There are also two receptors for endocannabinoids, CB1 and CB2. In the central nervous system, the majority of the attention is on CB1, which is found on presynaptic terminals and which serves as a break on the release of the amino acid transmitters, GABA and glutamate. The endocannabinoids themselves are synthesized in the postsynaptic terminal, and so this mechanism of feedback to the presynaptic terminal is an example of retrograde signaling and plays an important role in regulating synaptic transmission (Alger, 2002). The CB2 receptor was considered peripheral to the nervous system, being more important to the immune system, but recent work has highlighted a role for CB2 in the control of cell proliferation. Both anandamide and 2-AG bind to the CB1 and CB2 receptors, but with slightly differing affinities. The synthesis of the endocannabinoids is not as well understood as the active degradation, which is controlled by specific enzymes that when inhibited allow for an increase in 2-AG or anandamide in the synaptic cleft.

The female-biased higher rate of cell proliferation is eliminated when both sexes are treated with agonists that activate the CB2 receptor or block the degradation of the endogenous ligand. The sex difference is eliminated because the rate of proliferation in females is reduced, but there is no change in males. Indeed, a wide range of doses proved ineffective in males, suggesting there is a basal rate of proliferation in males and females that is impervious to regulation by endocannabinoids but that selectively in females that is an additional proliferative component that is stimulated by endocannabinoids. Females have lower levels of 2-AG and anandamide in the developing amygdala, which correlates with higher levels of the associated degradative enzymes, suggesting a lower overall tone of the endocannabinoid system. The selective sensitivity of females is a mystery, as there are few examples where only one sex is influenced by such a ubiquitous and dominant signaling pathways (Krebs-Kraft et al., 2010).

The second surprising component of the sex difference in cell proliferation in the developing amygdala is the ultimate fate of the new cells that are being regulated by endocannabinoids; they become glia, not neurons. Astrocytes are integral partners with neurons and critical regulators of both the formation and maintenance of synapses. A reasonable speculation is that the higher number of astrocytes in the female amygdala contributes to the sex difference in the synaptic profile found there, but other more subtle contributions cannot be ruled out. Regardless of what the role of the increased astrocytes might be, there is a correlation between that measure and a complex behavioral output, rough and tumble play. When play is quantified as the number and intensity of physical contacts made between individuals, males tend to outscore females by several fold. In rats, as well as other species, the sex difference in play is a direct result of androgen-mediated masculinization of the brain. If females are treated with androgen neonatally, they will play like males as juveniles (Meaney et al., 1983). The sexual differentiation of play is often touted as a purely androgen-mediated process, but recent evidence implicates aromatization as a contributing variable (Auger and Olesen, 2009; Auger et al., 2010). One of the aspects of play that is particularly interesting is its independence from hormones at the time the behavior is expressed. Unlike adult behaviors that require activation by steroids, play dominates during a period of the lifespan when circulating gonadal steroids range from low to undetectable. In other words, play is purely organized, with no activational component (Video 6).

Neither the neural circuitry nor the cellular mechanisms mediating the marked sex difference in play behavior is well understood, although progress is being made, including the surprising observation that inhibition of a repressive molecule, nuclear receptor corepressor (NCor), could increase play in males still further (Jessen et al., 2010). But other than hormones, there had been no agents identified that increase play in females, until the effects of endocannabinoids were explored. Treatment of neonatal females with anandamide increases their play behavior as juvenile to the level seen in males (Krebs-Kraft et al., 2010). This suggests that the greater number of astrocytes normally present in the female amygdala inhibits the expression of play, but a causal connection between these parameters has yet to be made.

VIDEO 6: Juvenile play behavior. http://tinyurl.com/mccarthy-video6.

Winged Messengers: Lessons from Birds and Flies

We have been emphasizing work in mammals, in particular rodents, but it is worth a brief segue into the world of other highly valuable animal models. Birds are of particular interest because of the robust nature of the sex difference in song, as well as sex differences in behavior. The rules governing the differentiation of these two separate processes appear to be distinct and are an active area of investigation. Fruit flies are attractive for their simplicity, their tremendous genetic tractability, and their easily quantified behavior, but as for hormones, not so much.

SEXUAL DIFFERENTIATION OF THE NEURAL CIRCUIT FOR SONG IN SONGBIRDS

Male songbirds sing a courtship song that females often cannot produce. The neural circuit for song is much larger in males than females, with pervasive sex differences in cellular morphology and organization. The earliest studies suggested that estradiol is responsible for the masculinization of the song system in males, predominantly because treating females with estradiol caused significant masculinization. However, treatments with gonadal steroids have not fully sex-reversed the song circuit in birds of either sex, and males have been relatively unaffected by various attempts to manipulate sex steroid hormones to prevent masculine development. Two major points are mentioned here and reviewed elsewhere (Wade and Arnold, 2004): (1) estrogen of neural (not gonadal) origin appears to be responsible for some parts of the masculinization process in males (Holloway and Clayton, 2001; London et al., 2006) and (2) studies of individual mutant birds suggests that the complement of sex chromosomes in brain cells may be the primary trigger of sexual differentiation (Figure 25). For example, genetically male tissue was found to be more masculine than genetically female brain tissue, although the tissues developed within the same animal and therefore in the same gonadal and hormonal environment (Agate et al., 2003). Sex differences in phenotype may result from constitutive sex differences in expression of sex chromosome genes, which are more widespread in birds because of ineffective dosage compensation of the sex chromosomes (Chen et al., 2005; Itoh et al., 2007).

Gynandomorphic Finch

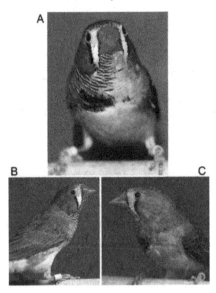

FIGURE 25: Gynandromorphic zebra finch. A gynandomorph is an individual that is genetically half male and half female, which can occur during fertilization and is more common, or at least more evident, in birds due to their sex-specific plumage. The study of one zebra finch allowed Arthur Arnold of UCLA to address questions of the relative importance of genes versus hormones in the sex differentiation of the song control circuitry and led him to conclude that genetics play a previously unappreciated important role (photo courtesy of Arthur Arnold).

COURTSHIP AND COPULATION IN *DROSOPHILA*

There are notable similarities and differences in the sexual differentiation of the *Drosophila* nervous system compared with mammals. Similarities include the existence of sex-typical behavioral patterns that are induced by pheromone signals emanating from the opposite sex. Likewise, development of the copulatory circuit in males is under control of a single male splice form of *fru* (Manoli et al., 2005), similar to the multilevel influence of a single sex-specific agent (testosterone) in male copulation in reptiles, birds, and mammals. However, unlike in mammalian models, there are distinct neural circuits in male versus female *Drosophila* that respond to a male-specific pheromone and elicit the appropriate behavioral response, i.e., inhibition of courtship of other males in males and induction of receptivity to males in females (Datta et al., 2008). The development of the male circuit is analogous to the SDN in rats in that it is the result of prevention of cell death, only in this case, it is due to the presence of the Fru protein, as opposed to estradiol exposure in mammals. Moreover, the induction of a sexually dimorphic neural circuit in *Drosophila* is cell autonomous, meaning that those neurons that express Fru are the ones that will constitute the dimorphic neural circuit (Kimura et al., 2005). In those instances in which we know the mechanism determining neuroanatomical sex differences in mammals, the opposite is true; there is a requirement for cell-to-cell communication and, in some instances, a clear demonstration that the sex differentiation process is not cell autonomous.

Sexual Differentiation of the Primate Brain

For many people, animals are interesting, but their primary interest is humans, are human brains sexually differentiated? Before we can address this question specifically, it is worthwhile to first review what we know about our next-of-kin, nonhuman primates, or monkeys. The rhesus macaque has been the primate model of choice because of its well-characterized reproductive physiology that bears a strong resemblance to humans and its ease of breeding and study in captivity, although the nature of the conditions of captivity has a profound effect on the emergence of sexually differentiated traits. Rhesus gestation lasts about 168 days (as opposed to closer to 300 in humans), and just as with humans, the testes differentiate very early in the process and begin synthesizing high levels of androgen. The peak androgen level in rhesus is approximately during midgestation, and this is followed by a second peak before birth and then elevated levels for a few months postnatally before declining (reviewed in Wallen, 2005). There are two aspects of hormonally mediated sexual differentiation of the rhesus brain that importantly distinguish it from the process in the rodent: (1) the sensitive period for organizational hormone actions is largely, if not exclusively, prenatal and (2) the hormone exerting the organizational effects is androgen, specifically testosterone, and not the metabolized product, estradiol. One of the costs of the sensitive period being prenatal is avoiding the confound of changes to the genitalia when considering the impact of hormones on sexual or even juvenile play behavior. Careful attendance to dose and timing confirms that elevated androgen during the latter part of gestation masculinizes male behavior independently of influencing the genitalia. However, this conclusion is restricted only to some aspects of male behavior, as there are complex effects of social rearing and experience, such as an early behavior called foot clasp mounting by infant males that presages adult behavior but requires cooperation from other males to perform. Thus, not surprisingly, primates are not as simple as rodents. One of the most robust behavioral sex differences in rhesus is interest in infants, with females showing substantially more interest than males, regardless of whether they have given birth to their own offspring, and the magnitude of the difference is even greater than that seen in humans. Manipulation of prenatal

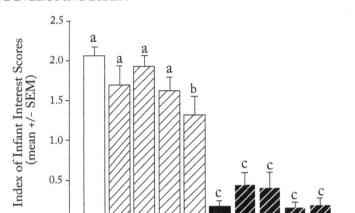

FIGURE 26: Interest in infants is sexually dimorphic in primates. By measuring the frequency and duration of interaction of juvenile rhesus monkeys with infants, the research group of Kim Wallen at the Yerkes National Primate Research Center generated an index of infant interest that revealed a marked sex difference, with females not surprisingly showing much greater interest in infants than males. Treatment of females with either androgen or the androgen receptor antagonist, flutamide, had relatively little effects in either males or females, with the exception of a modest effect of flutamide during late pregnancy on females. The authors conclude that the source of the robust sex difference in infants in rhesus monkeys remains unknown but is not likely due to socialization (reprinted with permission from Herman et al., 2003). Used with permission of Elsevier (Herman et al., 2003).

androgens has little to no effect on the propensity of rhesus females to interact with infants and precisely how this sex difference is established and maintained is not entirely clear (Herman et al., 2003) (Figure 26). Rhesus also show a sex difference in toy choice, a topic discussed further below, but consistent with humans, female rhesus are more interested in plush toys, while male rhesus are more inclined toward toys with moving parts, especially those that make noise (Figure 27).

FIGURE 27: Toy choice in primates is sexually dimorphic. Boys and girls are well-known to prefer different types of toys, but Melissa Hines has demonstrated that, even when they think no one is looking, boys tend to prefer toys that move and/or make noise while girls tend to prefer soft plush toys, especially stuffed animals or baby dolls. But we can never be sure how much parental coaching or other environmental factors influence toy choice in humans. This is not as likely to be a confound when studying animal models, such as nonhuman primates. When given a choice of toys to interact with, juvenile rhesus monkeys showed strikingly similar choices to that of humans. As seen here, a young female rhesus monkey has chosen a doll to play with, while the young male monkey is inspecting a toy car (photo courtesy of Kim Wallen).

Sexual Differentiation of the Human Brain

Recall that, in humans, we have two terms to denote the variable of maleness versus females, *sex* and/or *gender*. Gender encompasses the perception of sex by the individual and by society and recognizes that those perceptions impact on the individual in important biological ways. So, when we ask the question, is the human brain sexually differentiated, the answer is a qualified "yes," with the qualifications coming from our inability to make any causal statements since we cannot conduct controlled experiments to test specific hypotheses. We can do experiments in humans, but we cannot eliminate all the variables we want, and this is particularly important when looking at something as complex and intrinsic as the effect of sex or gender. However, we do not need to do experiments to be convinced that culture, society, environment, or the behavior of others toward us all have enormous influences on our own perception of our gender as well as how we behave. And, importantly, behavior itself is an important modifier of brain function, which makes it even more difficult to parse out the relative contributions of intrinsic hormones or genetics from external influences that then become internalized and ultimately biological. Moreover, the impact of external influences is often greatest in an immature or developing individual and may occur before the individual is even cognizant of the event. Here is an example, the so-called Baby X studies. Strangers were asked to interact with a newborn baby, which had either a pink or a blue cap on its head, regardless of what the actual sex of the child was. The behavior of the strangers toward the infants was markedly influenced by their perception of whether it was a boy (blue cap) or girl (pink cap) such that the girl was handled very gingerly, cooed at in a quiet voice, and told how cute or pretty she was, even though language to the baby at this point is a buzzing bumbling confusion When the same strangers interacted with the baby in the blue cap, their behavior was entirely different, vigorously picking the baby up, jostling a bit, and saying in a loud voice, "hey big guy, look at that big boy!" So it is right from the very start of life that we are barraged with signals that tell us who we are and what is important about us, which quite naturally then directs our behavior in an attempt to meet those expectations. This and other studies regarding human brain sex differences have been reviewed by leading experts in the field including Mellissa Hines (2002, 2004), Sheri Berenbaum (Bailey et al., 2002; Berenbaum, 1999), and Marc Breedlove (1994).

The barrage of external pressures to conform to specific gender roles continues throughout life and intensifies during the critical years of maturation, exerting enduring effects on the developing brain. Conflicts between our own desire to behave a certain way versus meet the external expectations of others can result in problems with gender identity and other forms of psychological turmoil. But not all external influences hammer an individual's phenotype into a specific gender role; some external variables are liberating, providing a means to broaden the repertoire of experiences of a child beyond that of their own gender. Gaining insight into the state of mind of an infant or child is challenging, and it is particularly challenging when the goal is to understand gender and avoid externally imposed expectations. Melissa Hines, a researcher at University of Cambridge in London, has been particularly adept at asking questions of children that address their brain gender and the interested reader is referred to her excellent book *Brain Gender* (Oxford University Press). One easily quantified readout of a child's internal brain gender is toy choice. Toys can be categorized as sex-typical for boys (i.e., cars, trains, tools, building blocks) or girls (i.e., dolls, kitchens, dress-up clothes). One generality between boys versus girls toys is a strong preponderance toward movement and noise in boys toys versus plushness and color in girls toys. Others toys are considered gender neutral, such as puzzles, drawing materials, or picture books. Hines and colleagues have used the well-established biases in toy choice to investigate the biological basis of gender and what influences it. To conduct the test, children are made comfortable in a room in which there is an array of toys to chose from and told to enjoy themselves freely for a specified period of time. In some tests, they are also told that at the end of the time, they can choose one toy to take home with them but that it will be their secret and only they will know which toy they choose. In the meantime, the children are observed from behind a two-way mirror by the adult experimenters who take note of the duration of time spent with each toy as well as the choice of the secret toy (which is not secret because the adults are watching and also the adults are smart enough to figure out which toy is missing at the end of the session). Quite predictably, boys will spend more time with male-typical toys and girls more time with female-typical toys. However, Hines has found that one of the strongest influences on toy choice outside of the gender expectation is whether the child has a sibling of the opposite sex. In other words, boys who have sisters are more likely to spend some time with female-typical toys while girls with brothers will spend time with boy-typical toys. Presumably, children with opposite sex siblings are not only exposed to toys favored by the opposite sex at home, but also observe other individuals enjoying those toys and the combination of exposure and behavioral modeling has a potent influence on their desire to either seek out those toys or in overcoming any concerns they might have of making an "improper" choice. But toy choice is not entirely malleable; if it were, it would not be a good readout of brain gender. Evidence of an intrinsic biological contribution is found in the study of a class of girls exposed to androgens during development, but we will return to this after a discussion of what is known about the biological contributions to sex differences in the human brain.

Sex differentiation of the human brain starts with the first principle: is there a similar pattern of differential steroid hormone exposure of males and females during gestation? In many ways, the answer is self-evident, as boys are born with a penis and scrotum, while girls are born with a clitoris and vagina, and the development of the genitalia are determined by hormones, androgens and estrogens. But there is also empirical evidence that, yes, indeed, humans have the same general pattern of hormone synthesis during development as nonhuman primates, which, in turn, are generally similar to the pattern seen across a wide range of mammals, including rodents. What we know about steroids in developing humans mostly comes from measures of umbilical cord blood and, more recently, amniotic fluid of women undergoing amniocentesis, which has become a routine procedure and thereby allows these samples to be considered a random representation of the healthy population.

A newborn baby boy has circulating testosterone levels at par with that of a 25-year-old adult (Corbier, 1990), and levels will remain elevated for 3 to 8 weeks before gradually declining to an undetectable level. The reemergence of androgens begins with steroidogenesis by the adrenal gland during a frequently ignored developmental stage called *adrenarche*, which begins in boys around 8 years old. Ultimately, androgen production by the testis beginning at the onset of true puberty, usually around 13 years old in boys, will swamp that of the adrenal gland. Adrenarche also occurs in girls and contributes to the development of axillary hair. Puberty in girls occurs several years earlier than boys and continues to accelerate in age of onset at an alarming rate of approximately 6 months a decade.

The recognized importance of steroid influences on brain development to childhood and adult behavior has led to a marked increase in attempts to correlate androgen levels with various outcomes, including the higher rates of aggression, anxiety, and even autism in males. These efforts have met with mixed success. There are instances where amniotic fluid levels of androgen are correlated with aspects of temperament, such as fear reactivity (Bergman et al., 2010), but other claims, such as the "extreme male brain" theory of autism (Baron-Cohen, 2002), are insufficiently supported at this time to be considered truly informative regarding the origins of this complex disorder. One of the major roadblocks to genuinely understanding the effects of steroids on the human brain is our inability to measure what the actual steroid levels are *in the brain*. Because we cannot extract chunks of brain, mash them up, and measure the steroid levels, we are left with a less-than-satisfactory approach of measuring the steroid levels in distant sites, such as the bloodstream or amniotic fluid. It is a bit like trying to understand what is coming out of the tap in a specific house in St. Louis, Missouri, by measuring the content of the Mississippi river flowing by several miles away. As we discussed early in this review, we now know steroids can be manufactured locally within the brain and that levels can vary substantially over time and between brain regions, and even in response to external challenges such as experience or changes in the environment. Thus, we are essentially blind to what the real levels of steroid are in a particular brain region at a particular time, and of course, the amount of steroid is only the beginning, there is also the amount of receptor, the cellular localization of the receptor, the availability of cofactors, and so on. So what is a neuroendocrinologist with an interest in humans to do? Turn to so-called natural experiments is the answer for many, as this has proven to be an approach with traction. Natural experiments are instances where genetic mutations, usually spontaneous, have led to either altered steroid production or loss of sensitivity to steroids. Thus, just as we manipulate steroids and steroid receptors in our animal models, there are humans with naturally occurring genetic mutations that essentially make them knockouts for a specific receptor or synthetic enzyme and we can assess aspects of their behavior just as we do in animals, but with all the caveats attendant to any study on humans.

ANDROGEN INSENSITIVITY SYNDROME

For some unknown reason, the AR is susceptible to spontaneous mutations, which either cripple it or render it completely unresponsive to steroid. The mutations include frame shifts, deletions, and insertions of stop codons and appear to cluster into hotspots (Figure 28). The vulnerability of the AR

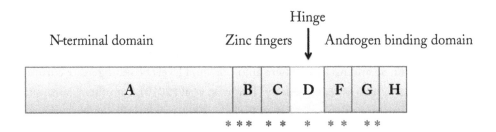

FIGURE 28: Androgen receptor. The androgen receptor consists of nine exons and is critical to sex determination by directing the formation of the reproductive tract, secondary sex characteristics, and masculinization of the brain in primates. Spontaneously arising mutations in the androgen receptor are not uncommon and tend to be clustered in the zinc finger, hinge and androgen-binding domains. Asterisk indicates mutations characterized in specific patients; those in red resulted in complete androgen insensitivity, whereas those in blue created partial androgen insensitivity (based on French et al., 1990).

occurs in other species as well, and strains of both rats and mice with spontaneously naturally mutated AR are available to researchers; these are called *tfm* mice (or rats) for testicularly feminized males, meaning they have testis but the external genitalia are female. The AR is located on the X chromosome, and mutations are therefore only of consequence to males, but females act as carriers, conferring an X chromosome with a mutated AR gene on their sons, while fathers would provide any daughters with at least one copy of an unmutated AR. In humans with complete androgen insensitivity, the genitals consist of a normal clitoris and a shorter than normal vagina that ends in a blind ended pouch, with no cervix, uterus, or fallopian tubes. Instead, the individual has internal undescended testis. Many individuals with complete androgen insensitivity used to go undetected and were not diagnosed until seeking an explanation for lack of menses or infertility. This happens less frequently now due to increased awareness and more careful neonatal examination. In either case, the principal response is removal of the testis, as they are prone to tumor development, followed by hormone replacement at the appropriate time. Even removal of the testis is becoming less frequent, as they continue to produce steroids and the cost/benefits of endogenous versus exogenous sources of steroid production are weighed and considered. Regardless, from the perspective of the brain, complete androgen insensitivity tells us that the psychosexual masculinization of the brain is driven by androgens and the AR, as these individuals identify as female with little to no gender identity conflict. Thus, the hormonal dependence of brain masculinization in humans is highly consistent with that of primates.

ESTROGEN RECEPTOR MUTATION AND AROMATASE DEFICIENCY

Clinical identification of androgen insensitivity syndrome has been established for decades, but awareness of a similar potential for spontaneous mutations in the ER or the aromatase enzyme has only appeared on the scene in the 1990s. Here is an interesting convergence of discoveries in man and mouse. One of the surprises generated by ERKOs, was the role of ER in bone maturation, in particular the conversion of the cartilage at the ends of long bones into bone to mark the end of growth. Turns out that the reason we stop growing shortly after puberty is that the increased estrogen production, in both girls and boys, completes the maturation of the bones. The later onset of puberty in boys in part contributes to their higher stature, and the relationship between cessation of growth and puberty is one of several reasons precocious puberty is a very bad thing. Conversely, if you either do not make estrogens or are blind to your own estrogens because of a mutated receptor, your bones do not know to stop growing and you just keep getting taller and taller. This was how in the late 1990s the first human with a mutated ER was identified—he went to the doctor because he just kept getting taller. Sequencing of his genome revealed a mutation that rendered his ER entirely insensitive to estrogen. Since that time, several other individuals with mutations that compromise their ability to synthesize estradiol have been identified for the same reason. Fortunately for them, there is an easy fix in hormone replacement therapy, unlike the unfortunate individual with the mutated ER who will just keep growing. From a psychosexual perspective, these individuals not surprisingly identify as males, but, although the sample size is exceedingly small, there are hints that aspects of their sex drive or interest in sex is lower than that of unaffected men, leaving the door open for a role for ER in masculinization of the human brain.

CONGENITAL ADRENAL HYPERPLASIA

The adrenal glands produce a substantial amount of steroid, particularly during development, and their principal role is the production of the mineralocorticoids and glucocorticoids, which are essential for normal lung maturation and kidney function. As noted in Figure 7, steroidogenesis is not a linear process; there are multiple intersecting pathways and enzymes that are common for some steroids and others that are unique. An analogy to a plumbing system is useful in considering that a clog in a key pipe in the network can cause a backup that increases flow out of the system via another drainage route. This is in essence what occurs in the condition called congenital adrenal hyperplasia (CAH), where debilitating mutations in key enzymes, usually 3β-dehydrogenase, stops up-flow toward the mineralocorticoids and glucocorticoids. thereby increasing flow toward androgen production. If this occurs in girls, the androgen production is often sufficient to result in virilization, meaning a partial masculinization of the genitalia, so that the clitoris may be markedly enlarged and the labia partially fused in an attempt to form a scrotum. This condition is often referred to as ambiguous genitalia, and the ambiguity is quickly tracked back to the elevated androgens during gestation. Whether male or female, these individuals require immediate medical attention, as there can be serious debilitating consequences of insufficient mineralocorticoid and glucocorticoid production, and some form of hormone replacement becomes part of their daily lives. But it is the girls that have been of great interest to neuroendocrinologists, as they represent an excellent opportunity to ask whether androgens masculinize the developing brain. As with any experiment, there are confounds, and that is certainly true here. For girls with CAH, we have to consider that many of them have had corrective surgery on their genitals and from a very early age, probably earlier than we even realize, have been aware that their gender identity is an issue of great interest, anxiety, and speculation by the adults around them. Moreover, the degree of individual variation in the amount of androgen exposure, combined with all the other variables inherent in human development, results in a highly heterogeneous population. Nonetheless, there have been important confirmations as well as refutations of the role of androgens in the psychosexual differentiation of humans based on observations of girls with CAH (see Puts et al., 2008). Most girls with CAH are heterosexual, just as are most unaffected girls, but in each population, there is a small percentage of girls that have a female sexual preference, and this small percentage is significantly larger in girls with CAH (Meyer-Bahlburg, 1982), suggesting androgens contribute to a sexual preference for females as partners, regardless of the sex of the individual. Independent of sexual orientation, girls with CAH exhibit a higher level of what most of us recognize as tomboyish behavior and what scientists categorize as "rough-and-tumble play." Play by juveniles is a universal across most any animal you can name that can move (sponges do not play), and in almost every instance, male play is characterized by a higher degree of intensity, meaning the interactions are more frequent and more physical. There is a great deal of individual variation and rough and tumble play, with many girls engaging in high levels while many normal males do not, but on average, when large numbers of individuals are assessed by self-questionnaire and parental questionnaire, males as a population are at the more extreme end in terms of frequency and intensity of rough play. On average, girls with CAH fall in the middle between unaffected females and males, although many girls with CAH are on the far end of female, while many unaffected females are indistinguishable from some of the more masculinized girls with CAH (Berenbaum, 1999). This variation may reflect naturally occurring variation in androgen exposure in all girls or may be the result of other variables such as the number of brothers a girl has, parental expectations, peer pressures, and who knows what else. Nonetheless, the consistently higher levels of rough-and-tumble play in girls with CAH is consistent with the interpretation that prenatal androgen acts on the brain to direct a masculine level of juvenile play.

Additional evidence for a masculinizing effect of prenatal androgen is also found in play, as we return to the example of toy choice. When the toy choices of 3- to 10-year-old children with CAH and their siblings were assessed, CAH girls displayed more male-typical toy choices than their unaffected sisters, whereas there was no difference between CAH boys and their brothers, both showing male-typical toy choices. The influence of parents was also assessed, and not surprisingly, mothers and fathers encouraged sex-typical toy choices, and this encouragement was particularly strong for girls with CAH choosing girl-typical toys. However, the desire to play with boy-typical toys in girls with CAH was not overridden by parental attempts to influence their choices (Pasterski et al., 2005).

The power of observations on play is the spontaneous expression of a child's interests, as opposed to testing of their skills or ability. A literal illustration of children's state of mind is found in a study of spontaneous drawing (Iijima et al., 2001). A group of Japanese children were provided a fresh box of crayons and a sheet of clean white paper and were asked to sit down and draw whatever came to mind. Analyses of both the color palette and aspects of the drawings revealed basic sex differences, with girls choosing warmer colors such as reds and yellows while boys tended toward cooler colors, lots of blues and blacks. The topic of choice also differed, with girls frequently drawing people, houses, flowers, and small animals, while boys preferred planes, trains, and automobiles. The girl drawings usually had a clear horizon, a line denoting the ground, a sun with rays in the corner, while the boy drawings frequently had no orientation, with frenetic activity in multiple dimensions. All of the male-biased attributes of drawings here were seen more frequently in drawings by girls with CAH, sometimes charmingly blending male and female characteristics such as one masterpiece of a very large fish swallowing whole an entire town (Figure 29).

Unaffected Girls and Boys

CAH girls

FIGURE 29: Children's drawings show gender bias that is influenced by prenatal androgen. Girls with CAH are exposed to androgens neonatally and are often studied to determine if their brains have been masculinized. Researchers in Japan asked children aged 3–5 years old to draw pictures of their favorite subjects and then analyzed them for content and color palette. In general, girls tended to draw pictures that included a horizon, a house, people, flowers, and a sun in the upper corner (two upper left drawings). Boys, on the other hand, had either no spatial orientation or tended to use a top-down perspective. Their subject matter was consistently of moving objects, planes, trains, and automobiles, and the color palette was dominated by blues and blacks. The drawings of CAH girls were either masculine or a blend of male and female characteristics, as exemplified in the bottom right drawing of a fish swallowing a town Used with permission of Elsevier (Iijima et al., 2001).

Together these studies paint a convincing picture of an impact of prenatal androgen exposure on the behavior of children, which we all know is charmingly different in little boys versus little girls. But the bigger question is whether prenatal androgen is determining adult differences in behavior, thereby directing sex differences in cognition and emotionality, i.e., does prenatal androgen make men better at math and women more likely to cry.

Overcoming the Hegemony of Hormones: Genes Matter Too

Up to this point, we have been in denial, as the discussion has focused exclusively on the rich and fascinating array of ways that steroids impact on the developing brain. But every brain cell has a sex, it is either XX or XY, and therefore, every cell in a male brain is to some degree fundamentally different than every cell in a female brain. The potential importance of this difference has been largely ignored for two reasons: (1) there did not seem to be a need, as hormones adequately explained everything, and (2) there were no tools by which to separate out the effects of sex chromosomes from sex hormones. Turns out neither of these are true, but it took the committed effort of one researcher to bring these facts to light. Arthur Arnold is a professor at UCLA who has been studying sex differences in the nervous system his entire career. A substantial portion of that career was spent studying birds, leading to many seminal discoveries, including the pivotal role of differential cell death as a regulator of sex differences in the size of particular subnuclei. Along the way, he began to observe a disturbing number of instances in which the organizational/activational paradigm did not fit the observations. At first, these were attributed to the exceptions that proved the rule or just anomalies associated with birds, but then a few tantalizing items turned up in the mammalian literature suggesting sex differences could emerge before the differentiation of the gonads, including differential gene expression in male and female brains (reviewed by Arnold and Burgoyne, 2004; Arnold et al., 2004). At the same time, scientists in England were developing a novel line of mice that would allow for the first time a separation of genetic and gonadal sex by moving the *Sry* gene, which codes for a testis, to an autosome and deleting it from the Y chromosome. This allowed for the generation of XY males that developed ovaries and XX females that developed testis. By comparing the XX females with testis to XX females with ovaries, you could assess the impact of the gonads independent of the Y chromosome, and by comparing XY males with testis to XX females with testis, you could assess the impact of the Y chromosome independent of the gonad. This has come to be known as the four-core genotype model (Figure 30) and has been extensively characterized to address these questions.

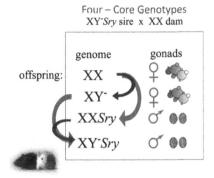

FIGURE 30: Mouse model for studying genetic versus hormonal effects. The generation of mice that have the *Sry* gene deleted from the Y chromosome but inserted into an autosome has allowed for the separation of genetic and hormonal effects. Animals that are XY-, meaning they have no *Sry*, develop ovaries and can be compared with XX animals that develop ovaries to distinguish genetic effects from gonadal effects; likewise, XY-*Sry* animals that do have a testis can be compared with XX Sry animals that also have a testis to again assess differences that are due to genes on the X or Y chromosome and not due to hormones from the gonads. Conversely, animals that are XX and have ovaries can be compared with animals that are XX Sry that have testis to control for genetics and vary the gonads, and the same for XY- and XY-Sry (figure courtesy of Arthur Arnold).

One of the principal findings from the study of the four-core genotype model is the hegemony of hormones, at least, that is the case when purely reproductive endpoints are considered, such as sex behavior and gonadotropin secretion and the brain regions controlling them. But when the investigation strays outside of the purely reproductive, numerous interesting effects emerge, beginning with the sex difference in vasopressin innervation of the lateral septum, which requires a combination of hormonal and genetic sex influences (De Vries et al., 2002). Behavioral analyses reveal a more central role for genetic sex in the control of aggression (Gatewood et al., 2006), habit formation, pain sensitivity, and alcohol preference (Barker et al., 2010). The brain structures mediating each of these has not been well characterized, but this use of the four-core genotype presents an excellent opportunity to make major progress in that arena.

The Y chromosome is a small vestige compared with the much larger X chromosome and has fewer genes, and this has presented mammals with the same problem facing any species in which sex is determined by chromosome matching—dosage compensation. Preventing the deleterious consequences of having a double dose of gene expression in one sex (female) versus the other (male) is achieved by silencing one gene, and this is done via X inactivation (Figure 31). The inactivation of one X chromosome is an epigenetic process (see below) of DNA methylation and histone modifications to prevent gene transcription. This is a tightly regulated process for the X chromosome and is initiated at the X-inactivation center by the noncoding RNA Xist. The inactivation process occurs as early as the blastocyst stage and is largely random whether the maternal or paternal X chromosome is shut off in a particular cell, but the decision is faithfully maintained in all subsequent cells derived from the original. However, some genes on the X chromosome escape inactivation, many more in the human than in the mouse as it turns out, and these usually have a homolog on the

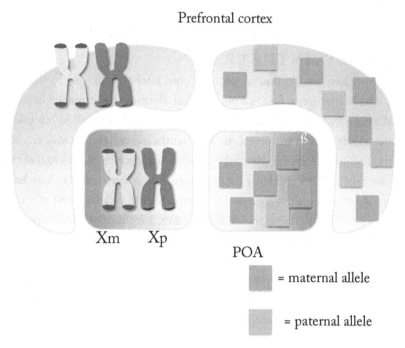

FIGURE 31: X chromosome inactivation and imprinting. Because females have two X chromosomes and males have only one, strategies have evolved to provide dosage compensation by silencing, or inactivating, one X chromosome in females. The inactivation process involves epigenetic changes such as methylation of the DNA. Males always inherit an X chromosome from their mothers, but females receive one from their mother and one from their father. In general, the inactivation is random throughout the body, but recent evidence indicates there can be systematic variation in which chromosome is inactivated in the brain, in particular the POA and prefrontal cortex where the maternal X tends to be more active than the paternal X. Genes on autosomes can also be subject to silencing, and if this is done in a systematic way so that the maternal versus paternal allele is preferentially silenced, it is referred to as imprinting. Many more genes are imprinted than previously believed, and there is a bias toward maternal allelic expression of genes in the POA, but this same bias was not observed in the prefrontal cortex (based on Gregg et al., 2010a). This is an exciting new area in the field of sex differences research and the significance of these biases in X inactivation and imprinting await further study.

Y chromosome that is similar to that on the X chromosome but may differ in important ways. Recent evidence suggests that the pattern of inactivation of the X chromosome is not random in the brain, with the maternal X being preferentially active in the preoptic area and prefrontal cortex of mice (Gregg et al., 2010a, b). A consequence of such a bias is a greater contribution from the maternal genome to her offspring, particularly in a brain region highly relevant to reproduction. Thus, there are many opportunities for X and Y genes to contribute to neural development and the establishment of sex differences in the brain.

EPIGENETICS AND THE DEVELOPMENT OF SEX DIFFERENCES IN THE BRAIN

Epigenetics literally means "in addition to genetics" and refers to changes to the DNA and associated proteins, which alter gene functioning without altering the sequence of nucleotide bases. Epigenetics is the basis by which all cell differentiation occurs. Every cell in our body contains our complete genome tightly packed within its nucleus, but the overwhelming majority of those 30,000 genes and intervening junk DNA are silenced by epigenetic processes to assure that the cells in your nose will stay nose cells and not suddenly become liver cells. More recently, there has been considerable excitement about epigenetics as a source of more subtle variation and as a means by which experience can have enduring effects on cell functioning, a form of cell memory. There is a range of opinions as to whether an epigenetic change is required to endure to the next generation. Purists believe the term *epigenetics* should be restricted to those changes that can be conferred through the germ cells to offspring in a Lamarckian form of inheritance of acquired characteristics. There are many examples of such epigenetic effects, including enduring influences of drug and toxin exposure, temperament, body fat, and nutritional status. Others believe that epigenetic changes can be far more plastic and come and go within a single individual within a single life span, and these are referred to as context dependent epigenetic changes (Crews, 2008). Both forms are mediated via the same mechanisms, making the argument more one of semantics and opinions on relative emphasis as opposed to a genuine scientific disagreement.

There are two main types of epigenetic changes: (1) those to the DNA and (2) those to associated histones, which are a component of the chromatin. Changes to the DNA itself are restricted to the introduction of a methyl group (CH_3) onto cytosine bases that sit just upstream of guanines and are usually found in greater abundance in the promoter region of specific genes. These are frequently referred to as CpG islands (Figure 32). The introduction of the methyl group has the potential to silence the gene by preventing access of the transcription complex; the methyl groups are rather bulky and simply keep the transcriptional proteins away from the DNA by steric hindrance, a fancy way of saying physically blocking them. Note the phrase "has the potential to silence genes," as not all methylation changes will in reality lead to reduced gene expression. Sometimes increased

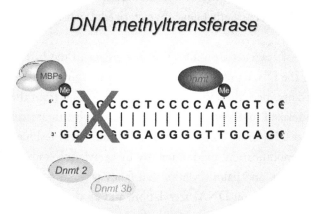

FIGURE 32: Epigenetics may underline long-term hormonal effects. Modifications to the DNA are done by the DNMT enzymes, for DNA methyl transferase, and consist of addition of a methyl group to cytosine residues that are located upstream (5′) of a guanidine residue. In general, this results in transcriptional repression, although views on the strictness of this are changing. There are several different DNMTs, some which are responsible for maintenance methylation and others respond to environmental and internal stimuli. Changes in methylation of DNA can also attract additional methyl binding proteins (MBPs), which further modulate transcription (figure courtesy of Jacklyn Schwarz).

methylation can actually increase gene expression, sometimes the effect is neutral, and sometimes there are likely to be effects that we do not really understand. The methylation of CpGs is achieved enzymatically by a family of DNA methyl transferases, referred to as DNMTs. DNMT1 is most important for routine maintenance methylation—the methylation that keeps your nose from turning into your liver. DNMT3a and 3b are the de novo methylators; they are responsible for changing the status of a particular genes expression profile in response to environmental challenges, experiences, and, as we will see, hormones. The addition of the methyl group to the cytosine involves a covalent bond, which is among the most stable in nature, and thus, methylation of a particular CpG is considered to be long-lasting and difficult to undo. However, increasing numbers of exceptions are beginning to undermine this view and raise the question of whether DNA methylation is a far more dynamic process than previously believed, particularly in the brain. In order for methylation to be truly dynamic, there must be a counterpart to active methylation, in other words, demethylation, and this would require specific enzymes as well. Candidate enzymes have been proposed, but the consensus remains too wobbly at this point to warrant discussion here. A second way that methylation can be undone is during the normal process of DNA repair, which goes on at a much higher rate than most of us are aware of but does not seem sufficient to explain the regulated and active degree of demethylation currently being characterized. It suffices to say that this is an area in which to stay tuned.

The second form of epigenetic modification is changes to the histones associated with the chromatin surrounding the DNA and participating in the packaging of DNA into nucleosomes. In order for genes to be transcribed, the DNA must first be unwound from the nucleosomes and the chromatin sufficiently relaxed to allow the transcription complex to assemble and associate with the appropriate sequences. Histones consist of multiple amino acid residues, and the peptide tails are a frequent target of modification, predominantly by acetylation and also by phosphorylation, glycosylation, ubiquitination, and palmitoylation (all fancy ways of adding stuff). Counter to the repressive effects of methylation on DNA, acetylation, and all the other -*lations*, mostly enhances gene transcription by promoting relaxation of the chromatin and enhancing assembly of the transcriptional complex. Also counter to the highly specific methylation of DNA, where changes to a single residue can have a major impact on gene expression, changes to histones tend to be more global and not as tightly associated with a particular gene, there is some degree of slippage between the surrounding histone cloud and the underlying DNA sequence. Acetylation and deacetylation are both active regulated processes driven by histone acetyl transferases (HATs) and HDACs (histone deacetylases).

Steroids and steroid receptors are particularly intriguing as potential context-dependent modulators of epigenetic changes because of the established association of these nuclear transcription factors with other cofactors that contain HAT or HDAC activity. Given that the effect of estrogens and androgens on the developing brain are enduring, the potential for an epigenetic component becomes even more likely. Demonstration that this is the case is found in a report that treatment of neonatal male rat pups with a drug that inhibits HDACs permanently disrupts the masculinization of sex behavior (Matsuda et al., submitted). There are also reports of sex differences in the amount of CpG methylation in the promoter region of ERα, ERβ, and PR in the neonatal rat, and estradiol itself regulates some but not all of these sex differences (see McCarthy et al., 2009). One intriguing aspect of this finding is that the pattern of DNA methylation seen within the first few days of life is not faithfully maintained into adulthood, but instead, new patterns appear that, while lawful in terms of reflecting a sex difference that is hormonally induced, are not the same as those seen within the early postnatal period. Moreover, the DNA methylation pattern observed does not correlate with changes in expression of any of the three genes examined, ERα, ERβ, and PR, suggesting that there is some additional form of cellular memory that steroids regulate during development. However, in other studies, a causal connection has been made between changes in CpG methylation of the ER promoter and ER expression approximately 10 days later (Kurian et al., 2010). In this instance, epigenetic changes serve as a convergence point for environment and hormones. As we noted before, the exaggerated attention that rat dams shower on their male offspring has important consequences to their development, and this now includes epigenetic marks on the ER promoter, which subsequently reduce ER expression. The same effect is achieved by a masculinizing dose of estradiol, demonstrating an autoregulatory loop in which, again, activated ER appears to impart restrictions on its own future expression and there is a beautiful redundancy in the experience of the male pup (i.e., more attention from mom) and his hormonal milieu (i.e., more estradiol). The study of epigenetic underpinnings to sexual differentiation of the brain is in its earliest stages but promises to be of profound interest. Given the overwhelming influence of experience and environment on sex and gender in humans, understanding how this responsive system modifies sex-specific parameters will be of central importance.

The Value of Understanding the Effect of Sex on the Developing Brain

That there are sex differences in the brain is self-evident, but how important they are or what the functional consequences of any difference is to the way males and females navigate the world is a matter of some debate. Moreover, the push to get scientists, in particular, neuroscientists, to incorporate both males and females into their experimental design continues to be an uphill battle. The reluctance to include males and females, much less to actively compare and contrast males and females, is likely due to multiple variables that range from a desire to save money by not incorporating twice as many animals (or people) in a study to the belief that any sex differences that do exist are trivial and not of importance to the particular endpoint under study. Also expressed on a regular basis is a desire to avoid confounds associated with the shifting hormonal cycles of females; in other words, males are just simpler. Many neuroscientists hold the belief that sex differences in the brain are largely limited to the neural control of reproductive behaviors, such as sexual and parenting behavior, and if one is instead studying learning and memory, there is no need to worry about the sex of the animals. But as should now be apparent from this review, sex differences are prevalent in pretty much every aspect of neural functioning, albeit to varying degrees.

References

Agate, R.J., Grisham, W., Wade, J., Mann, S., Wingfield, J., Schanen, C., Palotie, A., Arnold, A.P., 2003. Neural, not gonadal, origin of brain sex differences in a gynandromorphic finch. *Proc. Natl. Acad. Sci. U.S.A.* 100, pp. 4873–8.

Akerman, C.J., Cline, H.T., 2007. Refining the roles of GABAergic signaling during neural circuit formation. *Trends Neurosci.* 30, pp. 382–9.

Alger, B.E., 2002. Retrograde signaling in the regulation of synaptic transmission: Focus on endocannabinoids. *Prog. Neurobiol.* 68, pp. 247–86.

Amateau, S.K., McCarthy, M.M., 2004. Induction of PGE(2) by estradiol mediates developmental masculinization of sex behavior. *Nat. Neurosci.* 7, pp. 643–50.

Arnold, A.P., Burgoyne, P.S., 2004. Are XX and XY brain cells intrinsically different? *Trends Endocrinol. Metab.* 15, pp. 6–11.

Arnold, A.P., Xu, J., Grisham, W., Chen, X., Kim, Y.H., Itoh, Y., 2004. Minireview: Sex chromosomes and brain sexual differentiation. *Endocrinology.* 145, pp. 1057–62.

Auger, A.P., Jessen, H.M., Edelmann, M.N., 2010. Epigenetic organization of brain sex differences and juvenile social play behavior. *Horm. Behav.* [Epub ahead of print].

Auger, A.P., Olesen, K.M., 2009. Brain sex differences and the organisation of juvenile social play behaviour. *J. Neuroendocrinol.* 21, pp. 519–25.

Bailey, J.M., Bechtold, K.T., Berenbaum, S.A., 2002. Who are tomboys and why should we study them? *Arch. Sex. Behav.* 31, pp. 333–41.

Bakker, J., Baum, M.J., 2008. Role for estradiol in female-typical brain and behavioral sexual differentiation. *Front. Neuroendocrinol.* 29, pp. 1–16.

Bakker, J., De Mees, C., Douhard, Q., Balthazart, J., Gabant, P., Szpirer, J., Szpirer, C., 2006. Alpha-fetoprotein protects the developing female mouse brain from masculinization and defeminization by estrogens. *Nat. Neurosci.* 9, pp. 220–6.

Bakker, J., Honda, S., Harada, N., Balthazart, J., 2003. The aromatase knockout (ArKO) mouse provides new evidence that estrogens are required for the development of the female brain. *Ann. N.Y. Acad. Sci.* 1007, pp. 251–62.

Barker, J., Torregrossa, M., Arnold, A., JR, T., 2010. Dissociation of genetic and hormonal influences on sex differences in alcoholism-related behaviors. *J. Neurosci.* 30, pp. 9140–4.

Barna, B., Kuhnt, U., Siklos, L., 2001. Chloride distribution in the CA1 region of newborn and adult hippocampus by light microscopic histochemistry. *Histochem. Cell Biol.* 115, pp. 105–16.

Baron-Cohen, S., 2002. The extreme male brain theory of autism. *Trends Cogn. Sci.* 6, pp. 248–54.

Baum, M.J., Brand, T., Ooms, M.P., Vreeburg, J.T., Slob, A.K., 1988. Immediate postnatal rise in whole body androgen content in male rats: Correlation with increased testicular content and reduced body clearance of testosterone. *Biol. Reprod.* 38, pp. 980–6.

Becker, J.B., Arnold, A.P., Berkley, K.J., Blaustein, J.D., Eckel, L.A., Hampson, E., Herman, J.P., Marts, S., Sadee, W., Steiner, M., Taylor, J., Young, E., 2005. Strategies and methods for research on sex differences in brain and behavior. *Endocrinology.* 146, pp. 1650–73.

Berenbaum, S.A., 1999. Effects of early androgens on sex-typed activities and interests in adolescents with congenital adrenal hyperplasia. *Horm. Behav.* 35, pp. 102–10.

Bergman, K., Glover, V., Sarkar, P., Abbott, D.H., O'Connor, T.G., 2010. In utero cortisol and testosterone exposure and fear reactivity in infancy. *Horm. Behav.* 57, pp. 306–12.

Breedlove, S.M., 1994. Sexual differentiation of the human nervous system. *Annu. Rev. Psychol.* 45, pp. 389–418.

Breedlove, S.M.a.J., Cynthia L., 2001. The increasingly plastic, hormone-responsive adult brain. *Proc. Natl. Acad. Sci. U.S.A.* 98, pp. 2956–7.

Brennan, P.A., Keverne, E.B., 2004. Something in the air? New insights into mammalian pheromones. *Curr. Biol.* 14, pp. R81–9.

Burek, M.J., Nordeen, K.W., Nordeen, E.J., 1997. Sexually dimorphic neuron addition to an avian song-control region is not accounted for by sex differences in cell death. *J. Neurobiol.* 33, pp. 61–71.

Burks, S.R., Wright, C.L., McCarthy, M.M., 2007. Exploration of prostanoid receptor subtype regulating estradiol and prostaglandin E2 induction of spinophilin in developing preoptic area neurons. *Neuroscience.* 146, pp. 1117–27.

Cardona-Gomez, G.P., DonCarlos, L., Garcia-Segura, L.M., 2000. Insulin-like growth factor I receptors and estrogen receptors colocalize in female rat brain. *Neuroscience.* 99, pp. 751–60.

Cardona-Gomez, G.P., Mendez, P., DonCarlos, L.L., Azcoitia, I., Garcia-Segura, L.M., 2002. Interactions of estrogen and insulin-like growth factor-I in the brain: Molecular mechanisms and functional implications. *J. Steroid Biochem. Mol. Biol.* 83, pp. 211–7.

Chen, X., Agate, R.J., Itoh, Y., Arnold, A.P., 2005. Sexually dimorphic expression of trkB, a Z-linked gene, in early posthatch zebra finch brain. *Proc. Natl. Acad. Sci. U.S.A.* 102, pp. 7730–5.

Contreras, M.L., Wade, J., 1999. Interactions between nerve growth factor binding and estradiol in early development of the zebra finch telencephalon. *J. Neurobiol.* 40, pp. 149–57.

Cooke, B.M., Breedlove, S.M., Jordan, C.L., 1999. A brain sexual dimorphism controlled by adult circulating androgens. *Proc. Natl. Acad. Sci. U.S.A.* 96, pp. 7538–40.

Cooke, B.M., Stokas, M.R., Woolley, C.S., 2007. Morphological sex differences and laterality in the prepubertal medial amygdala. *J. Comp. Neurol.* 501, pp. 904–15.

Corbier, P., Dehennin, L., Castanier, M., Mebaza, A., Edwards, D.A., Roffi, J., 1990. Sex differences in serum luteinizing hormone and testosterone in the human neonate during the first few hours after birth. *J. Clin. Endocrinol. Metab.* 71, pp. 1344–8.

Corbier, P., Edwards, D.A., Roffi, J., 1992. The neonatal testosterone surge: A comparative study. *Arch. Int. Physiol. Biochim. Biophys.* 100, pp. 127–31.

Corbier, P., Roffi, J., Rhoda, J., 1983. Female sexual behavior in male rats: Effect of hour of castration at birth. *Physiol. Behav.* 30, pp. 613–616.

Crews, D., 2008. Epigenetics and its implications for behavioral neuroendocrinology. *Front. Neuroendocrinol.* 29, pp. 344–57.

Datta, S.R., Vasconcelos, M.L., Ruta, V., Luo, S., Wong, A., Demir, E., Flores, J., Balonze, K., Dickson, B.J., Axel, R., 2008. The *Drosophila* pheromone cVA activates a sexually dimorphic neural circuit. *Nature.* 452, pp. 473–7.

Davis, A.M., Ward, S.C., Selmanoff, M., Herbison, A.E., McCarthy, M.M., 1999. Developmental sex differences in amino acid neurotransmitter levels in hypothalamic and limbic areas of rat brain. *Neuroscience.* 90, pp. 1471–82.

Davis, E.C., Popper, P., Gorski, R.A., 1996. The role of apoptosis in sexual differentiation of the rat sexually dimorphic nucleus of the preoptic area. *Brain Res.* 734, pp. 10–18.

De Vries, G.J., Buijs, R.M., Van Leeuwen, F.W., 1984. Sex differences in vasopressin and other neurotransmitter systems in the brain. *Prog. Brain Res.* 61, pp. 185–203.

De Vries, G.J., Panzica, G.C., 2006. Sexual differentiation of central vasopressin and vasotocin systems in vertebrates: Different mechanisms, similar endpoints. *Neuroscience.* 138, pp. 947–55.

De Vries, G.J., Rissman, E.F., Simerly, R.B., Yang, L.Y., Scordalakes, E.M., Auger, C.J., Swain, A., Lovell-Badge, R., Burgoyne, P.S., Arnold, A.P., 2002. A model system for study of sex chromosome effects on sexually dimorphic neural and behavioral traits. *J. Neurosci.* 22, pp. 9005–14.

Demarque, M., Represa, A., Becq, H., Khalilov, I., Ben-Ari, Y., Aniksztejn, L., 2002. Paracrine intracellular communication by a Ca^{2+}- and SNARE-independent release of GABA and glutamate prior to synapse formation. *Neuron.* 36, pp. 1051–61.

Dittrich, F., Feng, Y., Metzdorf, R., Gahr, M., 1999. Estrogen-inducible, sex-specific expression of brain-derived neurotrophic factor mRNA in a forebrain song control nucleus of the juvenile zebra finch. *Proc. Natl. Acad. Sci. U.S.A.* 96, pp. 8241–6.

Forger, N.G., 2006. Cell death and sexual differentiation of the nervous system. *Neuroscience.* 138, pp. 929–38.

Forger, N.G., Rosen, G.J., Waters, E.M., Jacob, D., Simerly, R.B., De Vries, G.J., 2004. Deletion of *Bax* eliminates sex differences in the mouse forebrain. *Proc. Natl. Acad. Sci. U.S.A.* 101, pp. 13666–71.

Forger, N.G., Wong, V., Breedlove, S.M., 1995. Ciliary neurotrophic factor arrests muscle and motoneuron degeneration in androgen-insensitive rats. *J. Neurobiol.* 28, pp. 354–62.

Fusani, L., Metzdorf, R., Hutchison, J.B., Gahr, M., 2003. Aromatase inhibition affects testosterone-induced masculinization of song and the neural song system in female canaries. *J. Neurobiol.* 54, pp. 370–9.

Ganguly, K., Schinder, A.F., Wong, S.T., Poo, M., 2001. GABA itself promotes the developmental switch of neuronal GABAergic responses from excitation to inhibition. *Cell.* 105, pp. 521–32.

Garcia-Segura, L.M., Chowen, J.A., Duenas, M., Torres-Aleman, I., Naftolin, F., 1994. Gonadal steroids as promoters of neuro-glial plasticity. *Psychoneuroendocrinology.* 19, pp. 445–53.

Gatewood, J.D., Wills, A., Shetty, S., Xu, J., Arnold, A.P., Burgoyne, P.S., Rissman, E.F., 2006. Sex chromosome complement and gonadal sex influence aggressive and parental behaviors in mice. *J. Neurosci.* 26, pp. 2335–42.

Gorski, R.A., Harlan, R.E., Jacobson, C.D., Shryne, J.E., Southam, A.M., 1980. Evidence for the existence of a sexually dimorphic nucleus in the preoptic area of the rat. *J. Comp. Neurol.* 193, pp. 529–39.

Handa, R., Corbier, P., Shryne, J., Schoonmaker, J., Gorski, R., 1985. Differential effects of the perinatal steroid environment on three sexually dimorphic parameters of the rat brain. *Biol. Reprod.* 32, pp. 855–64.

Herman, R.A., Measday, M.A., Wallen, K., 2003. Sex differences in interest in infants in juvenile rhesus monkeys: Relationship to prenatal androgen. *Horm. Behav.* 43, pp. 573–83.

Hines, M., 2002. Sexual differentiation of human brain and behavior. In: *Hormones, Brain and Behavior*, Vol. 5, D. Pfaff, ed. Academic Press, London, UK, pp. 425–62.

Hines, M., 2004. *Brain Gender*, Oxford University Press, New York.

Hoffmann, C., 2000. COX-2 in brain and spinal cord implications for therapeutic use. *Curr. Med. Chem.* 7, pp. 1113–20.

Holloway, C.C., Clayton, D.F., 2001. Estrogen synthesis in the male brain triggers development of the avian song control pathway in vitro. *Nat. Neurosci.* 4, pp. 170–5.

Hutchison, J.B., Beyer, C., Hutchison, R.E., Wozniak, A., 1995. Sexual dimorphism in the development regulation of brain aromatase. *J. Steroid Biochem. Mol. Biol.* 53, pp. 307–13.

Ibanez, M.A., Gu, G., Simerly, R.B., 2001. Target-dependent sexual differentiation of a limbic-hypothalamic neural pathway. *J. Neurosci.* 21, pp. 5652–9.

Iijima, M., Arisaka, O., Minamoto, F., Arai, Y., 2001. Sex differences in children's free drawings: A study on girls with congenital adrenal hyperplasia. *Horm. Behav.* 40, pp. 99–104.

Ikeda, Y., Nishiyama, N., Saito, H., Katsuki, H., 1997. GABA-A receptor stimulation promotes survival of embryonic rat striatal neurons in culture. *Dev. Brain Res.* 98, pp. 253–8.

Isgor, C., Sengelaub, D.R., 1998. Prenatal gonadal steroids affect adult spatial behavior, CA1 and CA3 pyramidal cell morphology in rats. *Horm. Behav.* 34, pp. 183–98.

Itoh, Y., Melamed, E., Yang, X., Kampf, K., Wang, S., Yehya, N., Van Nas, A., Replogle, K., Band, M.R., Clayton, D.F., Schadt, E.E., Lusis, A.J., Arnold, A.P., 2007. Dosage compensation in birds versus mammals. *J. Biol.* 6, pp. 2.2–2.15.

Ivanova, T., Karolczak, M., Beyer, C., 2001. Estrogen stimulates the mitogen-activated protein kinase pathway in midbrain astroglia. *Brain Res.* 889, pp. 264–9.

Jessen, H.M., Kolodkin, M.H., Bychowski, M.E., Auger, C.J., Auger, A.P., 2010. The nuclear receptor corepressor has organizational effects within the developing amygdala on juvenile social play and anxiety-like behavior. *Endocrinology.* 151, pp. 1212–20.

Jordan, C.L., 1999. Glia as mediators of steroid hormone action on the nervous system: An overview. *J. Neurobiol.* 40, pp. 434–45.

Juntti, S.A., Coats, J.K., Shah, N.M., 2008. A genetic approach to dissect sexually dimorphic behaviors. *Horm. Behav.* 53, pp. 627–37.

Juntti, S.A., Tollkuhn, J., Wu, M.V., Fraser, E.J., Soderborg, T., Tan, S., Honda, S., Harada, N., Shah, N.M., 2010. The androgen receptor governs the execution, but not programming, of male sexual and territorial behaviors. *Neuron.* 66, pp. 260–72.

Kauffman, A.S., Park, J.H., McPhie-Lalmansingh, A.A., Gottsch, M.L., Bodo, C., Hohmann, J.G., Pavlova, M.N., Rohde, A.D., Clifton, D.K., Steiner, R.A., Rissman, E.F., 2007. The kisspeptin receptor GPR54 is required for sexual differentiation of the brain and behavior. *J. Neurosci.* 27, pp. 8826–35.

Kimchi, T., Xu, J., Dulac, C., 2007. A functional circuit underlying male sexual behavior in the female mouse brain. *Nature.* 448, pp. 1009–15.

Kimura, K., Ote, M., Tazawa, T., Yamamoto, D., 2005. Fruitless specifies sexually dimorphic neural circuitry in the *Drosophila* brain. *Nature.* 438, pp. 229–33.

Knoll, G.J., Wolkfe, C.A., Tobet, S.A., 2007. Estrogen modulates neuronal movements within the developing preoptic area–anterior hypothalamus. *Eur. J. Neurosci.* 26, pp. 1091–9.

Kudwa, A.E., Bodo, C., Gustafsson, J.A., Rissman, E.F., 2005. A previously uncharacterized role for estrogen receptor beta: Defeminization of male brain and behavior. *Proc. Natl. Acad. Sci. U.S.A.* 102, pp. 4608–12.

Kurian, J.R., Olesen, K.M., Auger, A.P., 2010. Sex differences in epigenetic regulation of the estrogen receptor-alpha promoter within the developing preoptic area. *Endocrinology*. 151, pp. 2297–305.

Leinekugel, X., Tseeb, V., Ben-Ari, Y., Bregestovski, P., 1995. Synaptic GABA$_A$ activation induces Ca^{2+} rise in pyramidal cells and interneurons from rat neonatal hippocampal slices. *J. Physiol.* 487, pp. 319–29.

Lephart, E.D., Simpson, E.R., McPhaul, M.J., Kilgore, M.W., Wilson, J.D., Ojeda, S.R., 1992. Brain aromatase cytochrome P-450 messenger RNA levels and enzyme activity during prenatal and perinatal development in the rat. *Brain Res. Mol. Brain Res.* 16, pp. 187–92.

Leypold, B.G., Yu, C.R., Leinders-Zufall, T., Kim, M.M., Zufall, F., Axel, R., 2002. Altered sexual and social behaviors in trp2 mutant mice. *Proc. Natl. Acad. Sci. U.S.A.* 99, pp. 6376–81.

Lieberburg, I., Krey, L.C., McEwen, B.S., 1979. Sex differences in serum testosterone and in exchangeable brain cell nuclear estradiol during the neonatal period. *Brain Res.* 178, pp. 207–12.

London, S.E., Monks, D.A., Wade, J., Schlinger, B.A., 2006. Widespread capacity for steroid synthesis in the avian brain and song system. *Endocrinology.* 147, pp. 5975–87.

Maclusky, N.J., Walters, M.J., Clark, A.S., Toran-Allerand, C.D., 1994. Aromatase in the cerebral cortex, hippocampus, and mid-brain: Ontogeny and developmental implications. *Mol. Cell. Neurosci.* 5, pp. 691–8.

Manoli, D.S., Foss, M., Villella, A., Taylor, B.J., Hall, J.C., Baker, B.S., 2005. Male-specific fruitless specifies the neural substrates of *Drosophila* courtship behaviour. *Nature.* 436, pp. 395–400.

Mathews, D., Edwards, D.A., 1977. Involvement of the ventromedial and anterior hypothalamic nuclei in the hormonal induction of receptivity in the female rat. *Physiol. Behav.* 19, pp. 319–26.

Matsumoto, A., Arai, Y., 1980. Sexual dimorphism in 'wiring pattern' in the hypothalamic arcuate nucleus and its modification by neonatal hormonal environment. *Brain Res.* 19, pp. 238–42.

McCarthy, M.M., Amateau, S.K., Mong, J.A., 2002a. Steroid modulation of astrocytes in the neonatal brain: Implications for adult reproductive function. *Biol. Reprod.* 67, pp. 691–8.

McCarthy, M.M., Auger, A.P., Bale, T.L., De Vries, G.J., Dunn, G.A., Forger, N.G., Murray, E.K., Nugent, B.M., Schwarz, J.M., Wilson, M.E., 2009. The epigenetics of sex differences in the brain. *J. Neurosci.* 29, pp. 12815–23.

McCarthy, M.M., Auger, A.P., Perrot-Sinal, T.S., 2002b. Getting excited about GABA and sex differences in the brain. *Trends Neurosci.* 25, pp. 307–12.

McCarthy, M.M., Konkle, A.T., 2005. When is a sex difference not a sex difference? *Front. Neuroendocrinol.* 26, pp. 85–102.

McEwen, B.S., 1981. Sexual differentation of the brain. *Nature.* 291; 610.

McEwen, B.S., Lieberburg, I., Chaptal, C., Krey, L.C., 1977. Aromatization: Important for sexual differentiation of the neonatal rat brain. *Horm. Behav.* 9, pp. 249–63.

Meaney, M.J., Stewart, J., Poulin, P., McEwen, B.S., 1983. Sexual differentiation of social play in rat pups is mediated by neonatal androgen-receptor system. *Neuroendocrinology*. 37, pp. 85–90.

Meyer-Bahlburg, H., 1982. Hormones and psychosexual differentiation: Implications for the management of intrasexuality, homosexuality and transsexuality. *Clin. Endocrinol. Metab*. 11, pp. 681–701.

Mong, J.A., Glaser, E., McCarthy, M.M., 1999. Gonadal steroids promote glial differentiation and alter neuronal morphology in the developing hypothalamus in a regionally specific manner. *J. Neurosci*. 19, pp. 1464–72.

Mong, J.A., Roberts, R.C., Kelly, J.J., McCarthy, M.M., 2001. Gonadal steroids reduce the density of axospinous synapses in the developing rat arcuate nucleus: An electron microscopy analysis. *J. Comp. Neurol*. 432, pp. 259–67.

Mong, J.A., Kurzweil, R.L., Davis, A.M., Rocca, M.S., McCarthy, M.M., 1996. Evidence for sexual differentiation of glia in rat brain. *Horm. Behav*. 30, pp. 553–62.

Mong, J.A., Nunez, J.L., McCarthy, M.M., 2002. GABA mediates steroid-induced astrocyte differentiation in the neonatal rat hypothalamus. *J. Neuroendocrinol*. 14, pp. 1–16.

Morris, J.A., Jordan, C.L., Breedlove, S.M., 2004. Sexual differentiation of the vertebrate nervous system. *Nat. Neurosci*. 7, pp. 1034–9.

Nordeen, E.J., Nordeen, K.W., 1996. Sex difference among nonneuronal cells precedes sexually dimorphic neuron growth and survival in an avian song control nucleus. *J. Neurobiol*. 30, pp. 531–42.

Nordeen, E.J., Nordeen, K.W., Sengelaub, D.R., Arnold, A.P., 1985. Androgens prevent normally occurring cell death in a sexually dimorphic spinal nucleus. *Science*. 229, pp. 671–3.

Nottebohm, F., Arnold, A.P., 1976. Sexual dimorphism in vocal control areas of the songbird brain. *Science*. 194, pp. 211–213.

Nunez, J.L., Bambrick, L.L., Krueger, B.K., McCarthy, M.M., 2005. Prolongation and enhancement of gamma-aminobutyric acid receptor mediated excitation by chronic treatment with estradiol in developing rat hippocampal neurons. *Eur. J. Neurosci*. 21, pp. 3251–61.

Nunez, J.L., McCarthy, M.M., 2009. Resting intracellular calcium concentration, depolarizing GABA and possible role of local estradiol synthesis in the developing male and female hippocampus. *Neuroscience*. 158, pp. 623–34.

Obrietan, K., van den Pol, A.N., 1995. GABA neurotransmission in the hypothalamus: Developmental reversal from Ca2+ elevating to depressing. *J. Neurosci*. 15, pp. 5065–77.

Ogawa, S., Chester, A.E., Hewitt, S.C., Walker, V.R., Gustafsson, J.A., Smithies, O., Korach, K.S., Pfaff, D.W., 2000. Abolition of male sexual behaviors in mice lacking estrogen receptors alpha and beta (alpha beta ERKO). *Proc. Natl. Acad. Sci. U.S.A*. 97, pp. 14737–41.

Ogawa, S., Lubahn, D.B., Korach, K.S., Pfaff, D.W., 1997. Behavioral effects of estrogen receptor gene disruption in male mice. *Proc. Natl. Acad. Sci. U.S.A.* 94, pp. 1476–81.

Ogawa, S., Washburn, T.F., Taylor, J., Lubahn, D.B., Korach, K.S., Pfaff, D.W., 1998. Modifications of testosterone-dependent behaviors by estrogen receptor—a gene disruption in male mice. *Endocrinology.* 139, pp. 5058–69.

Owens, D.F., Boyce, L.H., Davis, M.B.E., Kriefstein, A.R., 1996. Excitatory GABA responses in embryonic and neonatal cortical slices demonstrated by gramicidin perforated-patch recordings and calcium imaging. *J. Neurosci.* 16, pp. 6414–23.

Pankevich, D.E., Baum, M.J., Cherry, J.A., 2004. Olfactory sex discrimination persists, whereas the preference for urinary odorants from estrous females disappears in male mice after vomeronasal organ removal. *J. Neurosci.* 24, pp. 9451–7.

Pasterski, V., Geffner, M., Brain, C., Hindmwarsh, P., Brook, C., Hines, M., 2005. Prenatal hormones and postnatal socialization by parents as determinants of male-typical toy play in girls with congenital adrenal hyperplasia. *Child Dev.* 76, pp. 264–78.

Perrot-Sinal, T.S., Davis, A.M., Gregerson, K.A., Kao, J.P.Y., McCarthy, M.M., 2001. Estradiol enhances excitatory gamma-aminobutyric acid-mediated calcium signaling in neonatal hypothalamic neurons. *Endocrinology.* 143, pp. 2238–43.

Pfaff, D.W., 1966. Morphological changes in the brains of adult male rats after neonatal castration. *J. Endocrinology.* 36, pp. 415–416.

Pfaff, D.W., Sakuma, Y., 1979a. Facilitation of the lordosis reflex of female rats from the ventromedial nucleus of the hypothalamus. *J. Physiol.* 288, pp. 189–202.

Pfaff, D.W., Sakuma, Y., 1979b. Deficit in the lordosis reflex of female rats caused by lesions in the ventromedial nucleus of the hypothalamus. *J. Physiol.* 288, pp. 203–10.

Phoenix, C.H., Goy, R.W., Gerall, A.A., Young, W.C., 1959. Organizing action of prenatally administered testosterone proprionate on the tissues mediating mating behavior in the female guinea pig. *Endocrinology.* 65, pp. 369–82.

Plotkin, M.D., Kaplan, M.R., Peterson, L.N., Gullans, S.R., Herbert, S.C., Delpire, E., 1997a. Expression of the Na–K–2Cl cotransporter BSC2 in the nervous system. *Am. J. Physiol.* 272, pp. C173–C183.

Plotkin, M.D., Snyder, E.Y., Hebert, S.C., Delpire, E., 1997b. Expression of the Na–K–2Cl$^-$ cotransporter is developmentally regulated in postnatal rat brains: A possible mechanism underlying GABA's excitatory role in immature brain. *J. Neurobiol.* 33, pp. 781–95.

Puts, D.A., McDaniel, M.A., Jordan, C.L., Breedlove, S.M., 2008. Spatial ability and prenatal androgens: Meta-analyses of congenital adrenal hyperplasia and digit ratio (2D:4D) studies. *Arch. Sex. Behav.* 37, pp. 100–11.

Raisman, G., Field, P.M., 1971. Sexual dimorphism in the preoptic area of the rat. *Science.* 173, pp. 731–3.

Rhoda, J., Corbier, P., Roffi, J., 1984. Gonadal steroid concentrations in serum and hypothalamus of the rat at birth: Aromatization of testosterone to 17b-estradiol. *Endocrinology.* 114, pp. 1754–60.

Rico, B., Beggs, H.E., Schahin-Reed, D., Kimes, N., Schmidt, A., Reichardt, L.F., 2004. Control of axonal branching and synapse formation by focal adhesion kinase. *Nat. Neurosci.* 7, pp. 1059–69.

Rivera, C., Voipio, J., Payne, J.A., Ruusuvuori, E., Lahtinen, H., Lamsa, K., Pirvola, U., Saarma, M., Kaila, K., 1999. The K$^+$/Cl$^-$ co-transporter KCC2 renders GABA hyperpolarizing during neuronal maturation [see comments]. *Nature.* 397, pp. 251–5.

Roselli, C.E., Resko, J.A., 1993. Aromatase activity in the rat brain: Hormonal regulation and sex differences. *J. Steroid Biochem. Mol. Biol.* 44, pp. 499–508.

Schwarz, J.M., Liang, S.L., Thompson, S.M., McCarthy, M.M., 2008. Estradiol induces hypothalamic dendritic spines by enhancing glutamate release: A mechanism for organizational sex differences. *Neuron.* 58, pp. 584–98.

Schwarz, J.M., McCarthy, M.M., 2008. The role of neonatal NMDA receptor activation in defeminization and masculinization of sex behavior in the rat. *Horm. Behav.* 54, pp. 662–8.

Shughrue, P.J., Lane, M.V., Merchenthaler, I., 1997. Comparative distribution of estrogen receptor-alpha and -beta mRNA in the rat central nervous system. *J. Comp. Neurol.* 388, pp. 507–25.

Simerly, R.B., 2002. Wired for reproduction: Organization and development of sexually dimorphic circuits in the mammalian forebrain. *Annu. Rev. Neurosci.* 25, pp. 507–36.

Sinclair, A.H., Berta, P., Palmer, M.S., Hawkins, J.R., Griffiths, B.L., Smith, M.J., Foster, J.W., Frischauf, A.-M., Lovell-Badge, R., Goodfellow, P.N., 1990. A gene from the human sex determining region encodes a protein with homology to a conserved DNA-binding motif. *Nature.* 346, pp. 240–4.

Solum, D.T., Handa, R.J., 2002. Estrogen regulates the development of brain-derived neurtrophic factor protein in the rat hippocampus. *J. Neurosci.* 22, pp. 2650–9.

Speert, D.B., Konkle, A.T., Zup, S.L., Schwarz, J.M., Shiroor, C., Taylor, M.E., McCarthy, M.M., 2007. Focal adhesion kinase and paxillin: Novel regulators of brain sexual differentiation? *Endocrinology.* 148, pp. 3391–401.

Todd, B.J., Schwarz, J.M., McCarthy, M.M., 2005. Prostaglandin-E2: A point of divergence in estradiol-mediated sexual differentiation. *Horm. Behav.* 48, pp. 512–21.

Todd, B.J., Schwarz, J.M., Mong, J.A., McCarthy, M.M., 2007. Glutamate AMPA/kainate receptors, not GABAA receptors, mediate estradiol-induced sex differences in the hypothalamus. *Dev. Neurobiol.* 67, pp. 304–315.

Toran-Allerand, C.D., 2005. Estrogen and the brain: Beyond ER-alpha, ER-beta, and 17beta-estradiol. *Ann. N.Y. Acad. Sci.* 1052, pp. 136–44.

Wade, J., Arnold, A.P., 2004. Sexual differentiation of the zebra finch song system. *Ann. N.Y. Acad. Sci.* 1016, pp. 540–59.

Wallen, K., 2005. Hormonal influences on sexually differentiated behavior in nonhuman primates. *Front. Neuroendocrinol.* 26, pp. 7–26.

Webb, D.J., Donais, K., Whitmore, L.A., Thomas, S.M., Turner, C.E., Parsons, J.T., Horwitz, A.F., 2004. FAK–Src signalling through paxillin, ERK and MLCK regulates adhesion disassembly. *Nat. Cell Biol.* 6, pp. 154–61.

Weisz, J., Ward, I.L., 1980. Plasma testosterone and progesterone titers of pregnant rats, their male and female fetuses and neonatal offspring. *Endocrinology.* 106, pp. 306–13.

Wizemann, T.M., Pardu, M.-L., 2001. Exploring the biological contributions to human health. Does sex matter? Consensus Report. National Academy of Sciences, Washington, DC.

Woolley, C.S., 2007. Acute effects of estrogen on neuronal physiology. *Annu. Rev. Pharmacol. Toxicol.* 47, pp. 657–80.

Wright, C.L., McCarthy, M.M., 2009. Prostaglandin E2-induced masculinization of brain and behavior requires protein kinase A, AMPA/kainate, and metabotropic glutamate receptor signaling. *J. Neurosci.* 29, pp. 13274–82.

Zhang, J.-M., Konkle, A.T.M., Zup, S.L., McCarthy, M.M., 2008. Impact of sex and hormones on new cells in the developing rat hippocampus: A novel source of sex dimorphism? *Eur. J. Neurosci.* 27, pp. 791–800.

Classic References

1. Discovery of *Sry* (Sinclair et al., 1990).
2. Characterization of T in perinatal rat pups (Weisz and Ward, 1980).
3. Aromatization hypothesis (McEwen, 1981).
4. Organizational/activational hypothesis (Phoenix et al., 1959).
5. Sex differences in song control nuclei (Nottebohm and Arnold, 1975).
6. SDN of the POA (Gorski et al., 1980).
7. Cell death in the SNB (Nordeen at el., 1985).
8. First brain sex difference (Pfaff, 1966).
9. Second brain sex difference (Raisman and Fields, 1971).
10. Synaptic profile in the arcuate nucleus (Matsumoto and Arai, 1980).